Static and Dynamic Buckling of Thin-Walled Plate Structures

Tomasz Kubiak

Static and Dynamic Buckling of Thin-Walled Plate Structures

Springer

Tomasz Kubiak
Department of Strength of Materials
Lodz University of Technology
Lodz
Poland

ISBN 978-3-319-00653-6 ISBN 978-3-319-00654-3 (eBook)
DOI 10.1007/978-3-319-00654-3
Springer Cham Heidelberg New York Dordrecht London

Library of Congress Control Number: 2013939826

Printed on acid-free paper

Springer is part of Springer Science+Business Media (www.springer.com)

Contents

1 Introduction 1
1.1 Static Buckling and Postbuckling Behaviour 3
1.2 Dynamic Buckling 6
1.3 Literature Review 7
1.3.1 Buckling and Postbuckling Behaviour 7
1.3.2 Dynamic Buckling 12
References 16

2 Theory of Thin Plates for Laminates 27
2.1 Basic Assumptions 27
2.2 Geometrical Equations for Thin Plates 28
2.3 Constitutive Equations for Laminates 30
2.4 Generalized Sectional Forces 35
2.5 Dynamic Equations of Stability for Thin Orthotropic Plates . . . 40
References 46

3 Analytical–Numerical Method 47
3.1 Static Buckling and Postbuckling Behaviour 53
3.2 Natural Frequencies 60
3.3 Dynamic Buckling 63
References 64

4 Finite Element Method 67
4.1 Dynamic Buckling 68
4.2 Nonlinear Buckling Analysis 72
4.3 Linear Static Buckling Analysis 77
4.4 Modal Analysis 78
4.5 Element Type 79
4.6 Discretization 80
4.7 Load and Boundary Conditions 82
References 95

5 Dynamic Buckling Criteria 97
5.1 Volmir Criterion 98
5.2 Budiansky–Hutchinson Criterion 98
5.3 Ari–Gur and Simonetta's Criterion 99
5.4 Kleiber–Kotula–Saran Criterion 102
5.5 Author's Criterion 103
5.6 Teter Criteria 105
5.7 Petry–Fahlbusch Criterion 106
5.8 A New Approach to Dynamic Buckling Load Estimation 106
References 110

6 Thin Plates 113
6.1 Postbuckling Behaviour 118
6.2 Dynamic Buckling 122
References 138

7 Thin-Walled Columns 141
7.1 Local Buckling of Columns with Open Cross-Section 141
7.2 Interactive Buckling of Columns with Open Cross-Sections . . . 147
References 158

8 Girders, Beams Subjected to Pure Bending 159
8.1 Girders with Square Cross-Sections 159
8.2 Channel Section Beams 165
References 171

9 Thin-Walled Girders Subjected to Torsion 173
References 182

10 Thin-Walled Girders Subjected to Combined Load 183
References 188

Chapter 1
Introduction

The subject of this monograph is a study of buckling and postbuckling behaviour of thin plates and thin-walled structures with flat walls, subjected to static and dynamic load. The investigations have been carried out in the elastic range. The presented method of solutions as well as the exemplary results of calculations with some conclusions based on the conducted analysis have been the results of author's investigations carried out for the last 10 years.

Buckling and postbuckling behaviour of different structures subjected to static load are very well described in the worldwide literature. In the case of dynamic load – when the dynamic buckling problem is considered – there are numerous papers dealing with shells, some dealing with single plates made of different materials and with different boundary conditions, but still there is a lack of papers describing the behaviour of thin-walled complex structures consisting of flat plates. Some papers dealing with dynamic buckling of thin-walled plate structures have been written by scientists from the Lodz University of Technology [42, 43, 85, 92, 97–101, 108–119, 133–136] and the Lublin University of Technology [89, 106, 181]. This gap in the literature devoted to complex thin-walled structures was the main reason why the author of this monograph has decided to survey the problem of dynamic buckling of thin-walled plate structures.

The basic assumptions, a review of the thin plate theory, the methods used to determine the buckling load and a postbuckling analysis of thin-walled structures subjected to static and dynamic load are presented in this study. Two methods employed for static and dynamic buckling investigations are introduced. The ANSYS commercial software based on the finite element method and own analytical–numerical method developed for about twenty years in the Department of Strength of Materials of the Lodz University of Technology have been used.

The application of two different methods allows for wider understanding of the phenomenon. Two different methods can also enable one to uncouple the phenomena occurring at the same time and to attempt to estimate their impact on the final result.

A general mathematical model, adopted in the proposed analytical–numerical method, enables the consideration of all types of stability loss, i.e. local, global

T. Kubiak, *Static and Dynamic Buckling of Thin-Walled Plate Structures*,
DOI: 10.1007/978-3-319-00654-3_1,

(flexural, torsional, flexural–torsional, lateral and distortional) and interactive forms of buckling. The applied analytical–numerical method includes adjacent walls, a shear-lag phenomenon and a deplanation of cross-sections.

The chapter which discusses the finite element method presents both some theoretical as well as practical aspects that have been applied in resolving the issues of stability and dynamic buckling of thin-walled structures with flat walls. The ANSYS software [206], like other commercial programs based on the finite element method, is a closed code. It is almost impossible or very difficult to include own procedures in it. However, the ANSYS software allows one to prepare and include some user's routines, but generally this type of software is used for numerical experiments confirming the theoretical investigations or the results obtained using own software. It should be noted that a good interpretation of results of the analysis requires wide knowledge of the theoretical and analytical solution. In this publication, this is highlighted in chaps 6 to 9 by comparing the calculation results obtained with an application of both the aforementioned methods.

The thin-walled structure is a structure which consists of one or several thin plates or shells connected together at their common edges. Among thin-walled structures, plates, girders, beams, columns and shells are included. It is almost impossible to draw precisely the borderline between thin-walled elements and elements with average thickness. In the literature, one can find the information that the thin-walled rod is the one in which the wall thickness is at least 10 times smaller than the smallest cross-sectional dimension. This monograph presents computational examples of structures fulfilling the above definition of thin-walled structures.

Thin-walled structures have an ability to form freely the cross-section and, thus, to maximize mechanical properties of the material. Therefore, they have been more and more often used in many industries. Thin plates or thin-walled structures are used in sport and automotive industry, aerospace and civil engineering. As an example of such structural elements, a snowboard, a ski or poles can be mentioned, as well as all kinds of crane girders, structural components of automobiles (a car body sheathing and all longitudinal members), aircraft fuselages and wings, supporting structures of walls and roofs of large halls and warehouses.

As it is apparent from the above-mentioned applications, this type of structures can be made of isotropic materials (for example: steel, aluminium), as well as anisotropic and orthotropic ones (different kinds of composite materials, for example: a multilayered fibre composite, a sandwich composite, a functionally graded material, etc.). thin-walled structural elements have several advantages, such as:

- high dimensional accuracy;
- ease of installation;
- dimensional diversity;
- dimensional stability;
- relatively simple manufacturing technology;

- optimal distribution of the material in the cross-section due to the freedom of its formation;
- use of the mechanical properties of the material - lightweight and material savings;
- aesthetic appearance.

All the above structures, as well as many others which can be regarded as the thin-walled ones, exhaust their carrying capacity not by exceeding allowable stresses but by a stability loss. Therefore, not only the critical load but also the postbuckling behaviour of thin-walled structures subjected to static and dynamic load provides essential knowledge for designers. The use of more accurate mathematical models allows them to explore the phenomena occurring after a loss of stability and to describe more precisely their quick and easy software to be used to analyse the behaviour of thin-walled structures. Therefore, it has been decided to explore this issue, propose a mathematical model and a method of analysis of orthotropic thin-walled structures subjected to static and dynamic load.

1.1 Static Buckling and Postbuckling Behaviour

A stability loss or buckling is a system transition from one equilibrium state to another (the bifurcation point), or a jump from the stable to unstable equilibrium path (the limit point). The load resulting in a loss of stability is referred to as the critical load. The behaviour of the structure subjected to load higher than the critical one can be described by a stable (the growth of displacement is caused by increased load - see Fig. 1.1a) or unstable (displacements grow with a decrease in load) postbuckling equilibrium path (Fig. 1.1 b). Typical postbuckling equilibrium paths for such structures like columns (rods), plates, girders and shells are presented in Fig. 1.1.

The postbuckling behaviour of structures depends on their type. For example, cylindrical shells subjected to axial compression change their equilibrium stage (buckling) by the unstable bifurcation point or the limit point (Fig. 1.1e). Long rods or columns subjected to axial compression have usually sudden global buckling (the bifurcation point of the passage to a new postbuckling equilibrium path - Fig. 1.1d). Thin plates supported on all edges lose their stability having the local buckling mode and the stable postbuckling equilibrium path (Fig. 1.1d). The recalled types of buckling and postbuckling behaviour for given thin-walled structures are the same for ideal structures as for structures with geometrical imperfections. Columns made of thin prismatic plates can have the local buckling mode, the global (flexural, torsional or distortional) or coupled one.

In order to acquaint the reader with an application scope of the solution method and the results of calculations, a typical graph presenting the behaviour of thin plates or plated structures with a stable postbuckling equilibrium path is shown in Fig. 1.2.

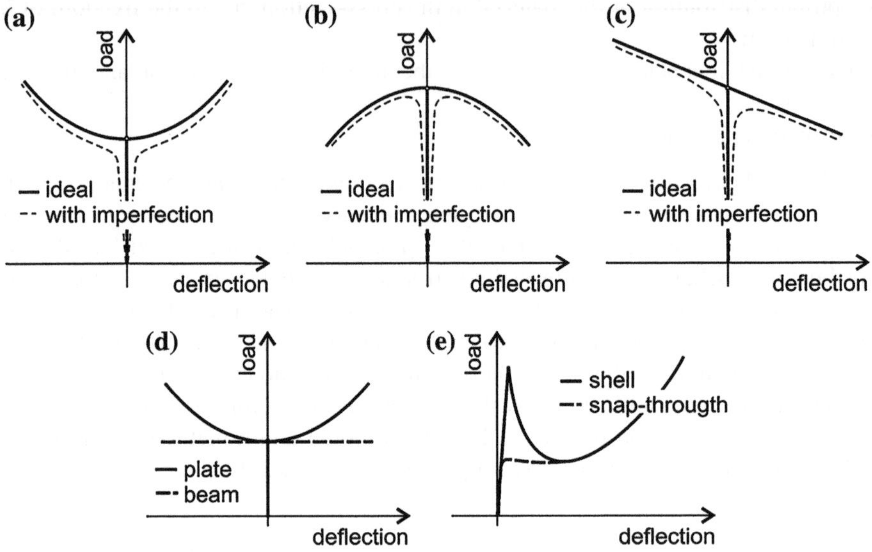

Fig. 1.1 Typical postbuckling equilibrium paths

Fig. 1.2 Diagram of behaviour of a thin-walled structure with a flat wall

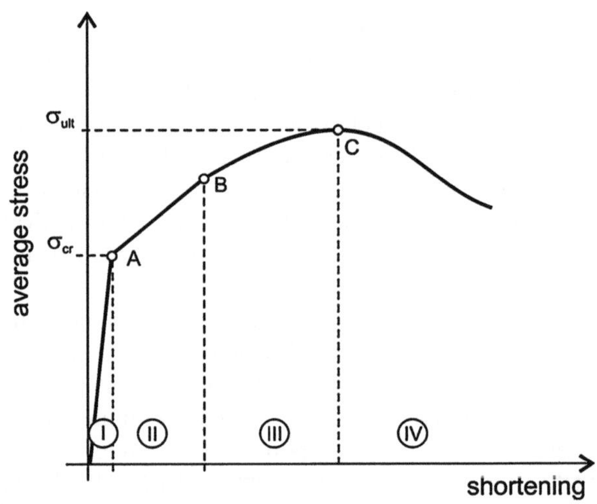

The first range (Fig. 1.2) is the prebuckling state. The structures after local buckling (point A – Fig. 1.2) are able to sustain further load (range II – Fig. 1.2) because the displacement increase is only possible by increasing the load value (stable postbuckling equilibrium path). A further increase in load leads to plasticity (point B – Fig. 1.2) or reaching a new, this time unstable, bifurcation point (global buckling). The range III is a postbuckling phase in the elastic–plastic range. The maximal load (point C – Fig. 1.2) is referred to as load carrying capacity after which the failure phase begins (range IV in Fig. 1.2).

The most dangerous form of a stability loss is the interactive buckling (coupled buckling) which usually causes the structure transition to the unstable equilibrium path, which leads to destruction of the structure with the load lower than the critical load corresponding to each mode separately. An interaction of different buckling modes occurs when the critical loads corresponding to different buckling modes are close to each other.

Generally, the buckling mode depends on the slenderness of compressed columns. It also depends on the type of structures, initial geometrical imperfections and also a type of load. Taking into account the length (parameter of the slenderness) of the compressed column, the buckling load can be divided into a local and global mode in the elastic and elastic–plastic range. Fig. 1.3 presents exemplary relations between buckling load and length for the compressed rod (Fig. 1.3 a) and the compressed thin-walled column (Fig. 1.3 b).

For a simple rod, the global buckling mode may be only taken into consideration. For a long and thin column, buckling takes place in the elastic range (slenderness of the rod is higher than the slenderness limit) and the buckling load is determined from the Euler formulae [22]. In the case when the rod is short, i.e., the rod slenderness is less than the limit of slenderness, buckling occurs in the elastic–plastic range and the buckling load may be determined using the well-known Jonson-Ostenfeld, Tetmajer-Jasinski or Rankine-Gordon formulae. For thin-walled columns, the situation is slightly more complicated – depending on length of the column, a different buckling mode may occur (Fig. 1.3b). For very short columns, the local buckling mode in the elastic–plastic range occurs (range I - Fig. 1.3b). For relatively short columns, different local buckling in the elastic range takes place (range II - Fig. 1.3b). For a long compressed column, a different global buckling mode (flexural, torsional, distortional) and their interaction with the local mode may occur (range III - Fig. 1.3b). Finally, for a very long column, the global mode takes place (range IV - Fig. 1.3b).

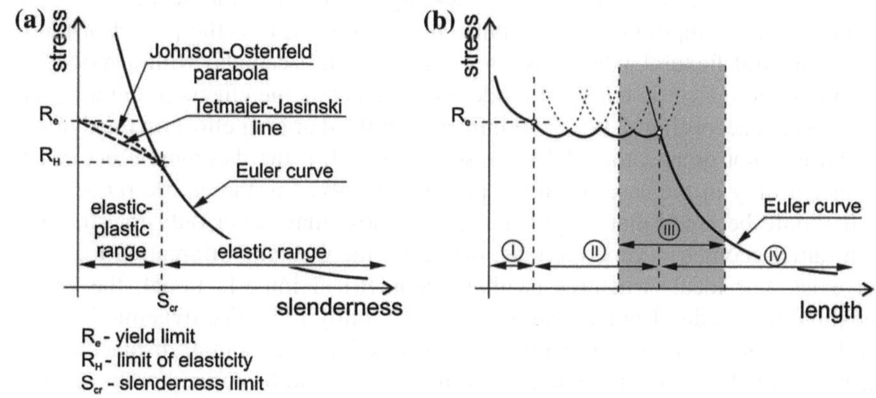

Fig. 1.3 Buckling load vs. slenderness or length of the compressed rod (a) and columns (b)

1.2 Dynamic Buckling

Real structures are not only subjected to static load but also to the dynamic one. The character of the phenomena that occur in the case of dynamic loads is determined by the duration of pulse load and its amplitude. In the literature, a quantity of "pulse intensity" [8] or "pulse velocity" [37] is introduced. In the case when the pulse duration is very short and the magnitude of load amplitude is relatively high, an impact phenomenon is observed. If the pulse duration corresponds to the period of natural vibrations and the magnitude of amplitude has an average value, the dynamic buckling occurs, whereas for a long period of the pulse duration, load is quasi-static.

It should be noted that during a short pulse duration, the dynamic critical load can be higher than the static buckling load.

In the literature, the dynamic buckling problem is analysed mainly for structures subjected to compression loads acting along the axis of the structure. Therefore, the present study has been limited only to the analysis of structures loaded in the plane of plates (walls of the beam-column under analysis), completely without the transverse load. Pulse loads are variable in time and act on the structure immediately, may be of a finite or infinite duration. A time diagram of dynamic loads with a finite duration (pulse loading) may take a parabolic, sinusoidal, rectangular, triangular, exponential or irregular shape [61]. There is a variety of pulses due to attempts to model the real load of the dynamic character. For example, pressure from a wave of the sea hitting a side of the ship or the boat is a sinusoidal pulse. The rectangular pulse models a hitting of bottom of the high-speed motor boat over the surface of water or a mass hitting the structure and then rebounded. Sudden and abrupt manoeuvres of flying objects generate dynamic loads of a trapezoidal shape. The exponential course describes the load caused by explosion, whereas the pulse with a triangular-step shape load describes load changes during a nuclear explosion. The above-mentioned examples have been confirmed by the experimental results which are presented, for example, in [63, 72, 196].

As mentioned above, the dynamic buckling occurs when the loading process is of intermediate amplitude and the pulse duration is close to the period of fundamental natural flexural vibrations (in a range of milliseconds) with a mode corresponding to the static buckling mode. In such a case, the effects of dumping can be neglected according to [96] – an influence of the damping effect on the dynamic response is not greater than 1 %. As shown in [133], the damping effect can be neglected only in the case when the problem is solved in the elastic range.

It should be noted that a dynamic stability loss may occur only for structures with initial geometric imperfections; therefore, the dynamic bifurcation load does not exist. For ideal structures (without geometrical imperfections), the critical buckling amplitude of pulse loading tends to infinity [25]. The dynamic buckling can be considered as strengthening the imperfections - the initial displacement of the structure. The critical dynamic buckling load should be defined on the basis of the assumed criterion.

The precise mathematical criteria were formulated for structures having the unstable postcritical equilibrium path or having the limit point [25, 168]. But for the structures having the stable postbuckling equilibrium path (thin plates, thin-walled beam-columns with the minimal critical load corresponding to local buckling), the precise mathematical criteria have not been defined till now. Therefore, Simitses [168] suggested not defining the dynamic buckling for structures with the stable postbuckling behaviour but rather defining it as a dynamic response to pulse loads. However, many scientists want to have a "tool" allowing them to define the critical amplitude of pulse load causing a loss of stability. Therefore, many authors adopted the criterion formulated for thin shells by Budiansky and Hutchinson [70]. Other scientists solving the dynamic buckling problem of thin plates proposed their own criteria. The oldest and probably the easiest to use is the dynamic stability criterion formulated by Volmir [195]. Other popular criteria are Ari-Gur and Simonetta criteria 8 and the failure criterion proposed by Petry and Fahlbush [147]. With the arrival of new papers [181] dealing with dynamic buckling, new criteria allowing to determine the critical amplitude of the pulse load have appeared - including a new criterion proposed by the author [115] of this publication (see Sect. 5.5).

In the analysis of the behaviour of the structure subjected to pulse loads, the concept of the dynamic load factor *DLF* defined as a ratio of the pulse amplitude to the static buckling load for the perfect structure is introduced. The critical value of the dynamic load factor DLF_{cr}, according to the above-mentioned reason, is determined on the basis of established criteria. The value of DLF_{cr} determines the ability of a structure to sustain dynamic loads. The author of this publication has proposed a new approach for determining the dynamic load factor [119] (see Sect. 5.8).

1.3 Literature Review

The literature overview presented below is focused only on main papers dealing with the elastic static buckling, the postbuckling behaviour and the dynamic buckling of thin-walled plates or structures composed of flat plate (walls). However, some of the most important publications dealing with other thin-walled structures, especially in the case of pulse load, are mentioned as well.

1.3.1 Buckling and Postbuckling Behaviour

Buckling and postbuckling of thin-walled structures subjected to static load have been investigated by many authors for more than one hundred years. The following scientists: Bernoulli and Euler [22], Timoshenko [186] and Volmir [194] should be included in the group of precursors of the investigations on stability of the thin-walled structure problem.

In the worldwide literature, there are numerous papers dealing with linear and nonlinear stability of thin-walled structures subjected to load of different kinds. Nowadays, numerous software packages usually based on the FEM, allowing to one calculate the critical load for most practical structures subjected to any type of load and to analyse their postbuckling behaviour, are available.

The widest development of research on stability of thin-walled isotropic structures took place in the 1970s and the 1980s. The exemplary papers dealing with local buckling of thin-walled structures are papers written by Davis and Hancock [44], Graves-Smith [58] or Mulligan and Pekoz [141]. Buckling and the postbuckling behaviour of isotropic thin-walled structures were analysed, for example, by Graves-Smith [59], Grimaldi and Pignataro [60], Koiter [78] Krolak [102]. Many works are devoted to multimodal (interactive) buckling and some of them are mentioned below. Koiter and Pignataro [80] presented a theoretical basis for the interaction of local and global buckling. Byskov and Hutchinson [29] dealt with the interactive buckling of cylindrical shells. An analysis of the interaction between the global mode and two local buckling modes was proposed by Koiter and van der Neut [81]. A more comprehensive review of the literature concerning the interactive buckling analysis of an isotropic structure can be found, for example, in Ali and Sridharan [5], Benito and Sridharan [21], Byskov [30], Koiter and Pignataro [79], Kolakowski [82–84], Manevich [132], Moellmann and Goltermann [139], Pignataro et al. [148], Pignataro and Luongo [149, 150], Sridharan and Ali [174, 175]. The interactive buckling of orthotropic structures has been discussed, for instance, in [89, 92].

In the wide world literature, works dealing with nonlinear problems of stability of thin-walled structures made of orthotropic materials can be found easily. The oldest work on this subject was published almost 80 years ago. Seydel [160], Smith [169] and Chang [34] dealt with orthotropic plate buckling. Reissner and Stavsky [152] published a study on the critical stress for anisotropic laminated plates with arbitrarily stacked layers. The theoretical background for buckling of composite and anisotropic plates was published by Lekhnitskii [120], Ambartsumyan [6], Ashton and Whitney [9] or Vinson and Chou [192]. In the literature, there are many works on anisotropic plates - among them, March's [137] and Thielmann's [185] works are worth mentioning. Fraser and Miller [53] established the critical load for orthotropic plates using the Ritz method. Mandell [131] presented the results of experimental studies on buckling of anisotropic rectangular plates with simply supported or clamped edges. Chailleux et al. [32] delivered the results of experimental studies on the stability of columns and square laminate plates. Noor [143] in his work presented a comparison between the classical theory of plates, the theory of linear shear and a 3-D theory for elastic stability of orthotropic laminated plates. Chandra and Raju [33] published a study on the postbuckling behaviour of orthotropic rectangular plates with simply supported edges. They analysed plates subjected to load causing uniform shortening of edges. They compared the results of their study with the previously published works.

A similar problem was solved by Prahakara and Chia [151]. They carried out a theoretical analysis of the postbuckling behaviour of orthotropic, rectangular plates with simply supported edges and subjected to biaxial compression. To describe the deformation, a double Fourier series was used. Massey [138] and Brunelle and Oyibo [24] looked for areas of instability for orthotropic plates subjected to pure shear. Libove [123] and Ting with co-authors [187] analysed the unstable behaviour of orthotropic plates under biaxial load.

In the 1980's and the 1990's, scientists widely used analytical–numerical and numerical methods to solve the stability problem. The finite strip method as well as the finite element method started to be used. Dawe and others [45, 46] analysed the postbuckling behaviour of geometrically nonlinear elastic plates and thin-walled prismatic layered composite laminates subjected to load causing uniform edges close-up. They used the finite strip method based on the classical plate theory and the first order deformation theory. Kasagi and Sridharan [75] used the finite strip method to study the stability and the postbuckling behaviour of multilayered composite plates subjected to shear. The authors employed a trigonometric function to describe deflection along the plate and assumed very long plates to decrease the boundary conditions influence. The finite element method was used by Hu and Tzeng [65] to analyse the stability of rectangular plates with elastic fibrous composite laminates with different arrangement of layers. The simply supported or clamped plates subjected to eccentric compressive load were analysed. They employed the commercial software – ABAQUS. Bao et al. [14] used the FEM to analyse the critical stress for flat rectangular orthotropic thin plates with different boundary conditions. Not only plates but also beam-columns made of anisotropic materials were investigated. Barbero and Tomblin [15] dealt with a loss of global stability of thin-walled I-section beam-columns made of various fibre composites. To determine the critical stress, the Southwell method was used. They compared the experimental results with the theoretical ones receiving a very good agreement – the percentage difference between both the methods was less than 6.2 %. Gupta and Rao [62] studied the stability of a thin cantilever beam with a Z-cross-section made of two (45/-45) or three- (0/45/0) layered laminates. The authors employed the finite element method and used two-node beam elements with three degrees of freedom at each node to build a discrete model of the beam under analysis.

In the last decade, Awrejcewicz and co-authors have published monographs [10–13] devoted to dynamics and statics of plates and shells made of iso- and orthotropic materials. They have presented a broad spectrum of analytical and numerical methods applied to solve problems of static stability and vibration of thin-walled structures.

Despite the fact that since the first work on stability of the compressed rod [22] and then other thin-walled structures, more than a century has passed, stability and load carrying capacity of thin-walled structures is still a current topic. Below there are quoted papers published during the last five years and dealing with stability and load carrying capacity of thin-walled structures.

Among the work of Polish authors on this subject, the papers published by Szymczak and Chroscielewski from the Gdansk University of Technology, Tomski and others from the Czestochowa University of Technology, Teter from the Lublin University of Technology, Garstecki, Magnucki and Zielnica from the Poznan University of Technology should be mentioned.

Szymczak [176] deals with stability of the construction of halls modelled as thin-walled frames. It has been proven that the obtained bifurcation point is unsymmetrical and unstable, which can lead to a reduction of critical loads due to some geometrical and loading imperfections.

Chroscielewski et al. in [36] discusses the effect of initial deflection on torsional buckling load of the thin-walled I-beam column. The numerical results obtained using the theory of thin-walled members and the non-linear 6-parameter theory of shells are compared. The authors have observed and analysed the localisation of local buckling modes.

Tomski et al. in their papers [188–190] present results of theoretical and numerical studies on the slender geometrically nonlinear system subjected to non-conservative loading.

Thin-walled beam-columns with intermediate stiffeners have been investigated by Teter and Kolakowski [182, 183]. They have analysed an interaction between global and local buckling and an influence of this interaction on buckling load.

The linear and non-linear stability analyses of double sigma members in the elastic range are analysed by Rzeszut and Garstecki [156, 157]. They use the finite element method to illustrate the importance of proper modelling of structures with slotted connections accounting for initial imperfections. The authors have also modelled the imperfections measured in situ [156] and have analysed their influence on the postbuckling behaviour of structures.

Magnucki with his team have published a few papers [128–130] devoted to global and local stability of cold-formed thin-walled channel beams with open or closed flanges. Papers [128, 130] present a simple analytical description and calculations, the numerical FEM analysis and the laboratory tests of selected beams. The numerical investigations of the optimization problem have been carried out by Magnucki and Paczos [129]. The authors have defined the optimization criterion and the dimensionless objective function as a quality measure.

The main area of Zielnica's interest are sandwich conical and cylindrical shells. In the latest papers, Zielnica et al. present a derivation of the stability equation and the method of solution for an elastic–plastic open conical shell made of orthotropic materials [145]. They take into consideration a bilayered open conical shell subjected to longitudinal force and lateral pressure. The solutions for a freely supported sandwich cylindrical shell with unsymmetrical faces, loaded by longitudinal forces, transversal pressure and shear, can be found in [73]. Paper [205] presents a buckling analysis and equilibrium stability paths of the sandwich conical shell with unsymmetrical faces subjected to combined load.

Kolakowski et al. [88, 90] have analysed the interactive buckling and the postbuckling behaviour of thin-walled columns with different cross-sections using the asymptotic Koiter theory for conservative systems. In [88], Kolakowski and

Kowal-Michalska analyse an influence of the axial extension mode on the interactive buckling of a thin-walled channel subjected to uniform compression.

Krolak with co-workers have performed experimental investigations and a numerical analysis of stability and load carrying capacity of multi-cell thin-walled columns of triangular and rectangular cross-sections [103–105]. The results of FEM calculations have been compared to the theoretical and experimental investigations.

Thin-walled beam-columns with open and closed cross-sections subjected to compression or pure bending have been analysed across the whole world.

Loughlan et al. have conducted a numerical analysis and experimental tests on lipped cross-section [125], I-section and box-section [124] struts. In the numerical analysis, the ANSYS software based on the finite element method has been employed. They have examined buckling, a postbuckled response and the failure mode of thin-walled struts assuming the elastic–plastic material behaviour. They have proposed FEM models and procedures for the determination of the coupled local-distortional interactive response of thin-walled lipped channel sections [125]. Ovesy et al. have employed the finite strip method [144] to carry out a numerical analysis and have compared the obtained results with the FEM and the experiments.

The postbuckling behaviour, the load carrying capacity estimation and the failure mode of stainless steel stub columns [126] and multilayered plate structures [94] have been analysed by Kotelko, Kowal-Michalska, Rhodes and others. Kotelko in cooperation with Dubina and others [191] has made an inventory and classified geometrical and analytical models for local-plastic mechanisms aiming to characterize the ultimate capacity of cold-formed steel sections.

A summary of the recent research on stability, postbuckling behaviour and load carrying capacity of cold-formed steel members and structures is presented in papers written by Rhodes and Macdonald [127, 153]. In [127] they discuss an influence of various aspects on the behaviour of thin-walled members under various loadings. The paper presents the investigation results carried out by students during their short duration programmes. Rhodes in his paper [153] presents the effects of end fixity on the plain channel column behaviour, the effects of transverse impact on struts and the damaged strut capacity and the large deflection behaviour of slender rings under diametrically opposed point loads. Paper [127] deals with cold-formed steel members and discusses the particular characteristics affecting their design.

The progress in computational methods has allowed for development and improvement of original programs useful for the buckling and postbuckling analysis of thin-walled structures. In recent years, two competitive software codes allowing the determination of critical load for uncoupled and coupled buckling have been developed. They enable also analysing the postbuckling behaviour of thin-walled beam-columns. One of them, called GBT [56], has been developed at the Technical University of Lisbon and is based on the generalized beam theory. The second one is called cFSM and is based on the constrained finite strip method. The cFSM software has been developed by Schafer [161] from Johns Hopkins University. The authors of the mentioned software in collaboration with each other

[2] and with other scientists have issued numerous publications explaining the behaviour of thin-walled beam-columns [3, 31, 48, 50, 57, 158, 166]. The developed software has been validated by a comparison of the calculation results with the experimental tests [51, 202] and other analyses using the commercial software based on the finite element method [3, 4, 31, 49, 50, 140, 166, 177].

For many years, the University of Sydney has been a recognized centre in which the experimental studies of long thin-walled beam-columns were carried out. Rasmussen et al. published the results of their experimental investigations on steel lipped channels [20, 155] and I-section columns [18] under axial compression. The interactions of local and overall buckling were considered. On the basis of the obtained test results, the FEM model incorporating the non-linear stress-strain behaviour, anisotropy, enhanced corner properties and initial imperfections was prepared and used in further investigations [17, 19]. The ABAQUS software was employed.

Rasmussen used also the isoparametric spline finite strip method to carry out an elastic buckling analysis of perforated thin-walled structures [52] and inelastic buckling of perforated plates [201]. In the case of study in the inelastic range, several material models were included in the analysis - elastic perfectly-plastic, linear hardening models and models with nonlinear yielding and isotropic hardening.

1.3.2 Dynamic Buckling

In the world literature, the topic of dynamic buckling has been known since the 1930s. The first known papers [93, 178] dealt with Euler buckling of columns subjected to compressive pulse with a finite duration. Those works did not take into account axial inertia forces. The authors noted that the rod took the dynamic load with amplitude higher than the static critical load. A further work carried out by Sevin [159] showed that the effect of axial inertia forces could be neglected as long as the column was in the elastic range.

A faster development of research and analysis of dynamic buckling dates at the 1960s and the early 1970s and is associated with the papers written by Budiansky [25–28, 70] dealing with shells and Volmir's publications [194, 195] devoted to the dynamic stability of thin plates and shells. In contrary to the articles published in the 1930s, the aforementioned works formulates the criteria allowing one to determine the critical value of amplitude of the dynamic load. The criterion allowing for determining the critical dynamic pressure was given by Budiansky and Roth [28]. They studied the response segment of a spherical shell loaded with pulse pressure. The critical pressure is defined as the one that causes a rapid change in strain. For pulse compression, a shell dynamic stability criterion was formulated by Hutchinson and Budiansky [70]. This work is often cited in the literature and concerns the dynamic stability of cylindrical shells with initial geometrical imperfections subjected to rectangular and triangular shape pulse load of a finite duration. The critical value of the amplitude of the pulse is based on the analysis of the system response to dynamic load, saying that the amplitude of pulse load causing unlimited growth of deflection is accepted as the critical load. Budiansky-Hutchinson criterion is adapted by many authors to determine critical loads of thin plates. It has been widely discussed in Sect. 5.2.

Volmir [194, 195] analysed simply supported rectangular plates and shells subjected to compressive pulse and shear pulse load. He proposed a very simple but time consuming method for dynamic buckling load determination (see Sect. 5.1).

The largest group of papers dealing with dynamic buckling is dedicated to shells. The most important are papers written by Budianky et al. [25–28, 70], Schokker, Sridharan and Kasagi [162], Huyan and Simitses [69], Sofiyev [170] Yaffe and Abramovich [200], Bisagni [23] and Virelli, Godoy and Suarez [193].

Another relatively large group of publications are papers related to thin columns with full- or thin-walled cross-sections, but taking into account only buckling in the Euler sense. Cui et al. [37–41, 64] were involved in experimental investigations. They tested columns with different dimensions of rectangular cross-sections and of various lengths. They assumed that the dynamic buckling occurred when the initial deflection and/or the initial speed of deflection were amplified to such a level that it began to grow to infinity. Zhanga and Taheri [204] analysed a dynamic response of composite beam-columns subjected to compression pulse load. The results of theoretical investigations and a comparison with experimental tests of rods impacted with high velocity are presented by Karagiozova and Jones [74]. Anwen and Wenying [7] used the finite strip method to analyse the dynamic buckling of thin rods in the elastic–plastic range. Kenny, Pegg and Taheri [76] used the finite difference method and the finite element method to analyse a dynamic response of rods subjected to pulse load with relative high pulse amplitude and short time of its duration. The results of studies on the influence of accidental geometrical imperfections on dynamic buckling are verified with the experimental tests [77].

The non-linear dynamic buckling analysis of the compressed rod modelled as a series of weightless rods connected one by one by a spring-damper system are presented by Kounadis, Gantes and Simitses [96]. For the assumed discrete model, the Lagrange equation was determined and solved. They defined the critical load as the one that leads to unrestricted movement or large jumps to remote stable equilibrium paths. The use of a damper in the model allowed for analysing the damping influence. The presented example showed that the difference in the results for the structure with viscous Rayleigh damping and without damping was only 0.2 %.

An isolated plate with different boundary conditions can be treated as a single wall of thin-walled structures. Knowing the behaviour of the plate, the local static and dynamic buckling load for beam-columns can be determined. Therefore, numerous publications dealing with the dynamic buckling of thin plates appear in the world literature. The most important are papers written by Volmir [194, 195], Weller, Abramovich and Yaffe [197], Abramovich and Grunwald [1], Ari-Gur and Simonetta [8], Petry and Fahlbush [147]. In the above-mentioned papers, the results of research and analysis of thin plates subjected to dynamic load are presented. An influence of the initial imperfection, different pulse shapes (sine, rectangle) and the pulse duration on a dynamic response of plates were presented. It was noted that dynamic buckling loads were usually greater than the static one but that relationship might be reversed for relatively large initial geometrical imperfections. Abramovich and Grunwald [1] conducted experimental investigations of composite plates with simply supported or clamped edges. Ari-Gur and Simonetta [8] used the finite difference method to solve the problem. Taking into account the inertial forces in the plane of the plate, they analysed the behaviour of isotropic and composite plates subjected to pulse load. On the basis of the results, four criteria of dynamic buckling were formulated (see Sect. 5.3). Taking into account the inertia forces in the plane of the plate, they analysed the behaviour of isotropic and composite plates subjected to pulse load. Petry and Fahlbusch [147] conducted a very comprehensive study examining an impact of the pulse duration to the maximum deflection, comparing different types of pulses – triangular, rectangular and sinusoidal. They also analyzed an influence of the size and the distribution (different number of halfwaves) of the geometry of the imperfection on the value of the dynamic critical load. On the basis of the results, a failure criterion was proposed, stating that the dynamic load is critical when at any point of the structure the equivalent stress was equal to the yield limit (see Sect. 5.7).

Cui, Cheong and Hao in their papers [35, 39] presented the results of experimental results for 15 rectangular plates clamped on loaded edges and with different boundary conditions (clamped - clamped, clamped - simply support, clamped - free edge) on longitudinal edges. All the above-mentioned plates were subjected to pulse load. The authors analysed different failure modes and proposed their own dynamic plastic failure criterion. They also performed a numerical analysis [40, 41] using the ABAQUS commercial software based on the finite element method. For the numerical analysis, they assumed an elastic–plastic model of the material with different coefficients of strengthening, noting that the dynamic critical load in the elastic–plastic range increased with an increasing material strengthening curve in the elastic–plastic state. Cui et al. studied an impact of the damping effect on the dynamic critical load, stating that the damping effect depended on the duration of the pulse and could be omitted when the pulse duration was close or less than the period of natural vibration of the structure under analysis.

The results of the dynamic buckling of the thin composite plate can be found in [43, 54, 97, 107, 108, 110, 146, 154].

In papers written by Papazouglou and Tsouvalis [146] and also Kowal-Michalska, Kolakowski and Czechowski [43, 98], the Galerkin method was used to derive the equation of motion, which was solved using the Runge–Kutta method.

Among the publications on dynamic buckling, there are papers which present an analysis of structures subjected to the simultaneous mechanical and thermal pulse load [55, 67, 68].

One- and bi-axial compression of rectangular plates was considered by Batra and Wei [16]. They studied the dynamic stability of orthotropic rectangular plates with elastic–plastic material models assuming the Hill's criterion of plasticity.

Plates with longitudinal stiffeners were the subject of paper [203] in which the authors present a theoretical analysis of the plates subjected to sinusoidal compression pulse with a duration equal to the period of fundamental natural vibrations. The problem was solved by an analytical–numerical method. To determine the dynamic buckling load, the Budiansky-Roth criterion was applied.

An extensive list of works dealing with dynamic buckling can be found, for example, in the book edited by Kowal-Michalska [99], written by Simitses [167] or Grybos [61]. Grybos' book [61] is the first synthetic scientific description dealing with the dynamic buckling of thin-walled structures subjected to pulse loading. The author presents the investigation results for bars, plates and shells.

Simitses in his work [168] presented the dynamic stability problems depending on the type of structure and its response to static loads. For the structures with "sudden" buckling (called snap-through buckling), he formulated concepts and methodologies to determine dynamic critical conditions. In the case of the structure with stable postbuckling equilibrium path, Simitses proposed that there was not any precise the critical dynamic buckling criterion. For the pulse load, we should rather speak about a dynamic response of the structure than about the dynamic buckling.

Among the wide literature dealing with the dynamic interactive buckling of columns (Euler buckling), only the papers written by Sridharan and Benito [173] and Kolakowski [85] have been found. Sridharan and Benito draw attention to the possibility of an interaction of local and global buckling. Kolakowski focuses on an interaction of various forms of global buckling. Other works on interactive buckling of thin-walled columns have been published by the author of this monograph [112, 113] and in cooperation with Kolakowski [91, 92].

The case of dynamic buckling in the elastic–plastic range, including the viscoplastic effect, has been investigated by Mania and Kowal-Michalska [133, 135, 136]. In [134] Mania has proven a significant impact of the strain rate effect on the dynamic buckling load of short columns. In the world literature, it is also possible to find a paper dealing with dynamic buckling of thin-walled structures subjected to combined load [106]. Czechowski [42] modelled a girder subjected to twisting and bending, considering only one plate subjected to shear and compression. A general summary showing which parameters have an influence on dynamic buckling of plated structures can be found in [100, 101].

Since the time a problem of dynamic stability started to be considered in the world literature, the majority of works has been devoted to shell structures and this trend is still present. During the last five years, numerous publications dealing with dynamic buckling of shells structures have been issued. Some of them to be mentioned are the papers written by Sofiyev with different co-authors [47, 142, 171, 172], Huang and Han [66] and Xu et al. [198, 199]. The latest papers written by Sofiyev are devoted to nonlinear dynamic buckling of cylindrical [172] and truncated conical shells [47, 142, 171] made of functionally graded materials. A similar problem, i.e., nonlinear dynamic buckling of composite cylindrical shells, made of ceramic–metal functionally graded materials, was discussed by Huang and Han [66]. Xu et al. in their papers present the local buckling problem of the cylindrical shell under torsion [199] and the local and global buckling problem of cylindrical shells under axial compression [198].

The theory allowing one to analyse dynamic axial-torsion buckling of structural frames was proposed by Leung [121, 122]. A nonlinear dynamic response of a sandwich plates was analyzed by Shariyat et al. [163–165]. They take into consideration the sandwich beam with SMA hybrid composite face sheets and a flexible core. Shariyat also analysed dynamic buckling of thin plates, paying attention to sandwich or composite structures. He used a finite element formulation based on a higher-order shear deformation theory of vibration and dynamic buckling of the FGM rectangular plate investigation. Paper [163] deals with plates with piezoelectric sensors and actuators subjected to thermo-electro-mechanical loads. The latest papers by Shariyat [164, 165] deal with dynamic buckling of imperfect viscoelastic sandwich plates. He analyses the nonlinear dynamic thermo-mechanical buckling and the postbuckling problem taking into consideration imperfect viscoelastic composite laminated/sandwich plates.

Currently two centres (i.e., the Lodz University of Technology and the Lublin University of Technology) are involved in the dynamic buckling problem of thin-walled structures with flat walls – with the plate model of beam-columns. Among the papers which have not been mentioned yet but have been published in the latest five years, there are publications by Teter [179, 180, 184], Kolakowski [86, 87], Kotelko and Mania [95] and Jankowski [71].

Kolakowski and Teter [86, 179, 180, 184] have investigated the interactive static and dynamic buckling of thin-walled columns with stiffeners subjected to axial compression. They have checked different dynamic buckling criteria, among them one proposed by Teter [179], which is based on phase-plane portraits.

Kotelko and Mania [95] have focused their attention on top hat section and plain channel section columns subjected to uniaxial uniform compression. They have made numerical and experimental investigations of an influence of loading velocity on the structural behaviour of TWCF members. The analysis has been performed using the FEM and the analytical solution based on the plastic yield-line analysis.

Jankowski [71] has used the finite element method software to analyse the dynamic buckling problem of thin-walled girders subjected to pulse loading. He has considered short columns (girders) with open (channel) and closed (rectangular and trapezoidal) cross-sections.

References

1. Abramovich H, Grunwald A (1995) Stability of axially impacted composite plates. Compos Struct 32:151–158
2. Adany S, Silvestre N, Schafer BW, Camotim D (2009) GBT and cFSM: Two modal approaches to the buckling analysis of unbranched thin-walled members. Adv Steel Constr 5(2):195–223
3. Adany S, Joo AL, Schafer BW (2010) Buckling mode identification of thin-walled members by using cFSM base functions. Thin-Walled Struct 48(10–11):806–817

4. Adany S, Visy D (2012) Global buckling of thin-walled simply supported columns: Numerical studies. Thin-Walled Struct 54:82–93
5. Ali MA, Sridharan S (1988) A versatile model for interactive buckling of columns and beam-columns. Int J Solids Struct 24(5):481–486
6. Ambartsumyan SA (1970) Theory of anisotropic plates. Technomic
7. Wang A, Tian W (2006) Development mechanism of local plastic buckling in bars subjected to axial impact. Int J Solids Struct 43:4578–4594
8. Ari-Gur J, Simonetta SR (1997) Dynamic pulse buckling of rectangular composite plates. Compos B 28B:301–308
9. Ashton JE, Whitney JM (1970) Theory of Laminated plates. Technomic
10. Awrejcewicz J, Krysko VA (2003) Nonclassical thermoelastic problems in nonlinear dynamics of shells. Springer-Verlag, Berlin
11. Awrejcewicz J, Andrianov IV, Manevitch LI (2004) Asymptotical mechanics of thin-walled structuresA. handbook. Springer-Verlag, Berlin
12. Awrejcewicz J, Krysko VA, Vakakis AF (2004) Nonlinear dynamics of continuos elastic sysytems. Springer-Verlag, Berlin
13. Awrejcewicz J, Krysko VA, Krysko AV (2007) Thermodynamics of plates and shells. Springer-Verlag, Berlin
14. Bao G, Jiang W, Roberts JC (1997) Analytic and finite element solutions for bending and buckling of orthotropic rectangular plates. Int J Solid Struct 34(14):1797–1822
15. Barbero E, Tomblin J (1993) Euler buckling of thin-walled composite columns. Thin-Walled Struct 17:237–258
16. Batra RC, Wei Z (2005) Dynamic buckling of a thin thermoviscoplastic rectangular plate. Thin-Walled Struct 43:273–290
17. Becque J, Rasmussen KJR (2009) Numerical investigation of the interaction of local and overall buckling of stainless steel I-columns. J Struct Eng 135(11):1349–1356
18. Becque J, Rasmussen KJR (2009) Experimental investigation of the interaction of local and overall buckling of stainless steel I-columns. J Struct Eng 135(11):1340–1348
19. Becque J, Rasmussen KJR (2009) A numerical investigation of local-overall interaction buckling of stainless steel lipped channel columns. J Constr Steel Res 65(8–9):1685–1693
20. Becque J, Rasmussen KJR (2009) Experimental investigation of local-overall interaction buckling of stainless steel lipped channel columns. J Constr Steel Res 65(8–9):1677–1684
21. Benito R, Sridharan S (1985) Mode Interaction in Thin-Walled Structural Members. J Struct Mech 12(4):517–542
22. Bernoulli J, Euler L (1910) Abhandlungen uber das Gleichegewicht und die Schwingungen der Ebenen Elastischen Kurven. Wilhelm Engelmann 175 Leipzig
23. Bisagni C (2005) Dynamic buckling of fiber composite shells under impulsive axial compression. Thin-Walled Struct 43:499–514
24. Brunelle EJ, Oyibo GA (1983) Generic buckling curves for specially orthotropic rectangular plates. AIAA J 21(8):1150–1156
25. Budiansky B (1965) Dynamic buckling of elastic structures: criteria and estimates. Report SM-7 NASA CR-66072
26. Budiansky B (1966) A survey of some buckling problem. Report SM-8 NASA CR-66071
27. Budiansky B, Hutchinson JW (1966) Dynamic buckling of imperfection-sensitive structures. In: Goetler H (ed) Proceedings of the Eleventh International Congress of Applied Mechanics, Munich pp 636–651
28. Budiansky B, Roth RS (1962) Axisymmetric dynamic buckling of clamped shallow spherical shells. Collected Papers on Instability of Shell Structures, NASA, TN-D-1510:597–606
29. Byskov E, Hutchinson JW (1977) Mode interaction in axially stiffened cylindrical shells. AIAA J 15(7):941–948
30. Byskov E (1988) Elastic Buckling Problem with Infinitely Many Local Modes. Mech Struct Mach 15(4):413–435
31. Camotim D, Basaglia C, Silvestre N (2010) GBT buckling analysis of thin-walled steel frames: A state-of-the-art report. Thin-Walled Struct 48(10–11):726–743

32. Chailleux A, Hans Y, Verchery G (1975) Experimental study of the buckling of laminated composite columns and plates. Int J Mech Sci 17:489–498
33. Chandra R, Raju B (1973) Postbuckling analysis for rectangular orthotropic plates. Int J of Mech Sci 16:81–97
34. Chang F (1958) Scientia Sinica VII 716
35. Cheong HK, Hao H, Cui S (2000) Experimental investigation of dynamic post-buckling characteristics of rectangular plates under fluid-solid slamming. Eng Struct 22:947–960
36. Chroscielewski J, Lubowiecka I, Szymczak C, Witkowski W (2006) On some aspects of torsional buckling of thin-walled I-beam columns. Comput Struct 84(29–30):1946–1957
37. Cui S, Hai H, Cheong HK (2001) Numerical analysis of dynamic buckling of rectangular plates subjected to intermediate velocity impact. Int J Impact Eng 25(2):147–167
38. Cui S, Hao H, Cheong HK (2001) Dynamic buckling and collapse of rectangular plates under intermediate velocity impact. In: Zaras J, Kowal-Michalska K, Rhodes J (eds) Proc. of Third International Conference of Thin-Walled Structures, Elsevier Science, Cracow pp 365–372
39. Cui S, Hao H, Cheong HK (2001) Dynamic buckling and post buckling of imperfect columns under fluid-solid interaction. Int J Solids Struct 38:8879–8897
40. Cui S, Hao H, Cheong HK (2001) Numerical analysis of dynamic buckling of rectangular plates subjected to intermediate-velocity impact. Int J Impact Eng 25:147–167
41. Cui S, Hao H, Cheng HK (2002) Theoretical study of dynamic elastic buckling of columns subjected to intermediate velocity impact loads. Int J of Mech Sci 44:687–702
42. Czechowski L (2008) Dynamic stability of rectangular orthotropic plates subjected to combined in-plane pulse loading in the elasto-plastic range. Mech Mech Eng 12(4):309–321
43. Czechowski L, Kowal-Michalska K (2006) Statecznosc dynamiczna ortotropowych plyt prostokatnych poddanych impulsowemu obciazeniu złozonemu. In: Proc. of Stability Structures XI-th Symposium, Zakopane
44. Davids AJ, Hancock GJ (1986) Compression tests of long welded I-section columns. J Struct Eng 112(10):2281–2297
45. Dawe DJ, Lam SSE, Azizian ZG (1993) Finite strip post-local-buckling analysis of composite prismatic plate structures. Comput Struct 48(6):1011–1023
46. Dawe DJ, Wang S (1994) Buckling of composite plates and plate structures using the spline finite strip method. Compos Eng 4(11):1089–1117
47. Deniz A, Sofiyev AH (2013) The nonlinear dynamic buckling response of functionally graded truncated conical shells. J Sound Vib 332(4):978–992
48. Dinis PB, Camotim D, Silvestre N (2010) On the local and global buckling behaviour of angle, T-section and cruciform thin-walled members. Thin-Walled Struct 48(10–11):786–797
49. Dinis PB, Camotim D (2011) Post-buckling behaviour and strength of cold-formed steel lipped channel columns experiencing distortional/global interaction. Comput Struct 89(3–4): 422–434
50. Dinis PB, Camotim D, Silvestre N (2012) On the mechanics of thin-walled angle column instability. Thin-Walled Struct 52:80–89
51. Dos Santos ES, Batista EM, Camotim D (2012) Experimental investigation concerning lipped channel columns undergoing local distortional global buckling mode interaction. Thin-Walled Struct 54:19–34
52. Eccher G, Rasmussen KJR, Zandonini R (2008) Elastic buckling analysis of perforated thin-walled structures by the isoparametric spline finite strip method. Thin-Walled Struct 46(2):165–191
53. Fraser HR Jr, Miller RE (1970) Bifurcation type of buckling of generally orthotropic clamped plates. AIAA J. 8(4):707–771
54. Gilat R, Aboudi J (1995) Dynamic buckling of nonlinear resin matrix composite structures. Compos Struct 32:81–88
55. Gilat R, Aboudi J (2002) The Lyapunov exponents as a quantitive criterion for the dynamic buckling of composite plates. Int J Solids Struct 39:467–481

56. Goncalves R, Dinis PB, Camotim D (2009) GBT formulation to analyse the first-order and buckling behaviour of thin-walled members with arbitrary cross-sections. Thin-Walled Struct 47(5):583–600
57. Goncalves R, Ritto-Correa M, Camotim D (2010) A new approach to the calculation of cross-section deformation modes in the framework of generalized beam theory. Comput Mech 46(5):759–781
58. Graves-Smith TR (1969) The local buckling of box girders under bending stresses. Int J Mech Sci 11:603–612
59. Graves-Smith TR (1972) The postbuckled behaviour of a thin-walled box beam in pure bending. Int J Mech Sci 14:711–722
60. Grimaldi A, Pignataro M (1979) Post-buckling behaviour of thin-walled open cross-section compression members. J Struct Mech 7(2):143–159
61. Grybos R (1980) Statecznosc konstrukcji pod obciazeniem uderzeniowym. PWN
62. Gupta RK, Rao KP (1985) Instability of laminated composite thin-walled open-section beams. Compos Mater 4:299–313
63. Hall AW, Sawyer RH, McKay JM (1958) Study of ground-reaction forces measured during landing impact of a large airplane. NACA Tech. Rep. 4247
64. Hao H, Cheong HK, Cui S (2000) Analysis of imperfect column buckling under intermediate velocity impact. Int J Solids Struct 37:5297–5313
65. Hsuan-Teh Hu, Wen-Long Tzeng (1997) Buckling analysis of skew composite laminate plates. In: XI International Conference On Composite Materials, Gold Coast, Queensland Australia
66. Huang H, Han Q (2010) Nonlinear dynamic buckling of functionally graded cylindrical shells subjected to time-dependent axial load. Compos Struct 92(2):593–598
67. Chen Hui, Virgin LN (2006) Finite element analysis of post-buckling dynamics in plates - Part I: An asymptotic approach. Int J Solids Struct 43:3983–4007
68. Chen Hui, Virgin LN (2006) Finite element analysis of post-buckling dynamics in plates - Part II: A non-stationary analysis. Int J Solids Struct 43:4008–4027
69. Huyan X, Simitses GJ (1997) Dynamic buckling of imperfect cylindrical shells under axial compression and bending moment. AIAA J 35(8):1404–1412
70. Hutchinson JW, Budiansky B (1966) Dynamic buckling estimates. AIAA J 4–3:525–530
71. Jankowski J (2008) Dynamic response of thin-walled composite beam-columns with closed and open cross-sections. Mech Mech Eng 12(3):255–265
72. Jansen O, Langseth M, Hopperstand OS (2004) Experimental investigations on the behaviour of short to long square aluminium tubes subjected to axial loading. Int J Impact Eng 30:973–1003
73. Jaskula L, Zielnica J (2011) Large displacement stability analysis of elasticplastic unsymmetrical sandwich cylindrical shells. Thin-Walled Struct 49(5):611–617
74. Karagiozova D, Jones N (1996) Multi-degrees of freedom model for dynamic buckling of an elastic–plastic structure. Int J Solids Struct 33(23):3377–3398
75. Kasagi A, Sridharan S (1992) Postbuckling analysis of layered composites using p-version finite strips. Int J Numer Meth Eng 33:2091–2107
76. Kenny S, Pegg N, Taheri F (2000) Dynamic elastic buckling of a slender beam with geometric imperfections subject to an axial impulse. Finite Elem Anal Des 35:227–246
77. Kenny S, Taheri F, Pegg N (2002) Experimental investigations on the dynamic plastic buckling of a slender beam subject to axial impact. Int J Impact Eng 27:1–17
78. Koiter WT (1963) Elastic stability and post-buckling behaviour. In: Proceedings of the Symposium on Non-linear Problems, Univ. of Wisconsin Press, Wisconsin pp 257–275
79. Koiter WT, Pignataro M (1974) An alternative approach to the interaction between local and overall buckling in stiffened panels. In: Buckling of structures Proceedings of IUTAM Symposium, Cambridge pp 133–148
80. Koiter WT, Pignataro M (1976) A general theory for the interaction between local and overall buckling of stiffened panels. WTHD, Report No. 556, Delft University

81. Koiter WT, van der Neut A (1980) Interaction between local and overall buckling of stiffened compression panels, part I :51–56 and part II :66–86. In: Rhodes J, Walker AG (eds) Thin-Walled Structures. Granada, St.Albans, London
82. Kolakowski Z (1989) Some thoughts on mode interaction in thin-walled columns under uniform compression. Thin Wall Struct 7(1):23–35
83. Kolakowski Z (1993) Interactive buckling of thin-walled column-beams with open and closed cross-sections. Thin-Walled Struct 15(3):159–183
84. Kolakowski Z (1993) Influence of modification of boundary conditions on load carrying capacity in thin-walled columns in the second order approximation. Int J Solid Struct 30(19):2597–2609
85. Kolakowski Z (2007) Some aspects of dynamic interactive buckling of composite columns. Thin-Walled Struct 45(10–11):866–871
86. Kolakowski Z (2009) Static and dynamic interactive buckling of composite columns. J Theor App Mech 47(1):177–192
87. Kołakowski Z (2010) Static and dynamic interactive buckling regarding axial extension mode of thin-walled channel. J Theor Appl Mech 48(3):703–714
88. Kolakowski Z, Kowal-Michalska K (2011) Interactive buckling regarding the axial extension mode of a thin-walled channel under uniform compression in the first nonlinear approximation. Int J Solids Struct 48(1):119–125
89. Kolakowski Z, Kowal-Michalska K (eds) (2012) Static, dynamic and stability of structures, Vol. 2: Statics, dynamics and stability of structural elements and systems, A series of monographs, Technical University of Lodz Press
90. Kolakowski Z, Krolak M (2006) Modal coupled instabilities of thin-walled composite plate and shell structures. Compos Struct 76:303–313
91. Kolakowski Z, Kubiak T (2004) Multiple interaction of dynamic buckling modes in thin-walled members subjected in-plane pulse loading. In: Proceedings of 4th International Conference on Coupled Instabilities in Metal Structure, Rome, Italy, 27–29 September 2004
92. Kolakowski Z, Kubiak T (2007) Interactive dynamic buckling of orthotropic thin-walled channels subjected to in-plane pulse loading. Compos Struct 81:222–232
93. Koning C, Taub J (1934) Impact buckling of thin bars in the elastic range hinged at both ends. NACA TM 748
94. Kotelko M, Kowal-Michalska K, Kubiak T, Kolakowski Z, Gradzki R (2008) Estimation of load-carrying capacity of multi-layered plated structures. Thin-Walled Struct 46(7–9):1003–1010
95. Kotelko M, Mania RJ (2012) Quasi-static and dynamic axial crushing of TWCF open-section members. Thin-Walled Struct 61:115–120
96. Kounadis AN, Gantes C, Simitses G (1997) Nonlinear dynamic buckling of multi-dof structural dissipative system under impact loading. Int J Impact Eng 19(1):63–80
97. Kowal-Michalska K, Kolakowski Z, Mania R (2004) Estimation of dynamic load factor for orthotropic plate subjected to in-plane pulse loading. In: Proc. of Fourth Inter. Conf. on Thin-Walled Structures, Loughborough
98. Kowal-Michalska K, Czechowski L, Kolakowski Z (2006) Dynamic buckling of rectangular plates subjected to combined in-plane loading. In: Proc. of International Colloquium on Stability and Ductility of Steel Structures, IST Press, Lisbon, Portugal
99. Kowal-Michalska K (ed) (2007) Statecznosc dynamiczna kompozytowych konstrukcji plytowych. WNT, Warsaw
100. Kowal-Michalska K (2010) About somc important parameters in dynamic buckling analysis of plated structures subjected to pulse loading. Mech Mech Eng 14:269–279
101. Kowal-Michalska K, Mania JR (2008) Some aspects of dynamic buckling of plates under in-plane pulse loading. Mech Mech Eng 12(2):135–146
102. Krolak M (ed) (1995) Statecznosc, stany zakrytyczne i nosnosc graniczna cienkosciennych dzwigarow o scianach plaskich. Series of monographs, Technical University of Lodz Press

103. Krolak M, Kowal-Michalska K, Mania RJ, Swiniarski J (2007) Experimental tests of stability and load carrying capacity of compressed thin-walled multi-cell columns of triangular cross-section. Thin-Walled Struct 45(10–11):883–887
104. Krolak M, Kowal-Michalska K, Mania RJ, Świniarski J (2009) Stability and load carrying capacity of multi-cell thin-walled columns of rectangular cross-sections. J Theor Appl Mech 47(1):435–456
105. Krolak M, Mania RJ (2010) Critical and postcritical behavior of thin walled multicell column of open profile. Mech Mech Eng 14(2):281–290
106. Krolak M, Mania RJ (eds) (2011) Static, dynamic and stability of structures. Vol. 1: Stability of thin-walled plate structures. A series of monographs, Technical University of Lodz Press
107. Kumar Ravi L, Datta PK, Prabhakara DL (2003) Dynamic instability of laminated composite plates subjected to partial follower edge load with damping. Int J Mech Sci 45:1429–1448
108. Kubiak T (2005) Dynamic buckling of thin-walled composite plates with varying widthwise material properties. Int J of Solid Struct 45:5555–5567
109. Kubiak T (2005) Dynamic buckling of thin-walled girders with channel cross-section. VDI-Ber 1899:69–78
110. Kubiak T (2005) Dynamic buckling of thin composite plates. In: Proceedings IUTAM Symposium on Multiscale Modelling and Fracture Processes in Composite Materials, Springer pp 123–130
111. Kubiak T (2005) Wyboczenie dynamiczne cienkosciennych dzwigarow kompozytowych. In: Proc. of IX-th Scientific-Technical Conference "FEM software in computer aided analysis, design and manufacturing", Gizycko, Poland pp 89–96
112. Kubiak T (2006) Interactive dynamic buckling of thin-walled girder with channel cross-section. In: Proceedings of International Colloquium on Stability and Ductility of Steel Structures, IST Press, Lisbon, Portugal
113. Kubiak T (2006) Interakcyjne wyboczenie cienkosciennych slupow kompozytowych poddanych obciazeniu impulsowemu. In: Proc. of Stability of Structure XI-th Symposium, Zakopane, Poland
114. Kubiak T (2006) Kryterium statecznosci dla konstrukcji cienkosciennych obciazonych impulsowo. In: Proc. of Stability of Structure XI-th Symposium, Zakopane, Poland
115. Kubiak T (2007) Criteria for dynamic buckling estimation of thin-walled structures. Thin-Walled Struct 45(10–11):888–892
116. Kubiak T (2007) Interakcyjne wyboczenie dynamiczne cienkosciennych slupow. Technical University Press, Lodz
117. Kubiak T (2007) Metoda elementow skonczonych jako eksperyment numeryczny statecznosci dzwigarow cienkosciennych obciazonych statycznie i dynamicznie. In: Niezgoda T (ed) Numerical Analysis of selected problem in mechanics, Military University of Technology Press, Warsaw pp 209–228
118. Kubiak T (2011) Estimation of Dynamic Buckling for Composite Columns with Open Cross-Section. Comput Struct 89(21–22):2001–2009
119. Kubiak T, Kowal-Michalska K (2012) A new approach to dynamic buckling load estimation for plate structures. In: Proceedings of Stability of Structures 13th Symphosium, Zakopane, Poland pp 397–406
120. Lekhnitskii S (1947) Anisotropnyje plastinki. Moscow-Leningrad
121. Leung AYT (2008) Exact dynamic stiffness for axial-torsional buckling of structural frames. Thin-Walled Struct 46(1):1–10
122. Leung AYT (2009) Nonconservative dynamic axial–torsional buckling of structural frames using power series. Int J Mech Sci 51(11–12):807–815
123. Libove C (1983) Buckle pattern of biaxially compressed simply supported orthotropic plates. Int J Compos Mater 17:45–48

124. Loughlan J, Yidris N, Cunningham PR (2011) The effects of local buckling and material yielding on the axial stiffness and failure of uniformly compressed I-section and box-section struts. Thin-Walled Struct 49(2):264–279
125. Loughlan J, Yidris N, Jones K (2012) The failure of thin-walled lipped channel compression members due to coupled local-distortional interactions and material yielding. Thin-Walled Struct 61:14–21
126. Macdonald M, Rhodes J, Kotelko M (2007) Stainless steel stub columns subject to combined bending and axial loading. Thin-Walled Struct 45(10–11):893–897
127. Macdonald M, Heiyantuduwa MA, Rhodes J (2008) Recent developments in the design of cold-formed steel members and structures. Thin-Walled Struct 46(7–9):1047–1053
128. Magnucka-Blandzi E, Magnucki K (2011) Buckling and optimal design of cold-formed thin-walled beams.Review of selected problems. Thin-Walled Struct 49(5):554–561
129. Magnucki K, Paczos P (2009) Theoretical shape optimization of cold-formed thin-walled channel beams with drop flanges in pure bending. J Constr Steel Res 65(8–9):1731–1737
130. Magnucki K, Paczos P, Kasprzak J (2010) Elastic buckling of cold-formed thin-walled channel beams with drop flanges. J Struct Eng 136(7):886–896
131. Mandell JM (1968) An experimental study of the postbuckling of anisotropic plates. MSc thesis, Case Western Reserve University
132. Manevich AI (1988) Sviazannaja poteria ustoicivosti srzatoi podkreplennoi paneli. Mekh Tverd Tela 23(5):152–159
133. Mania RJ (2011) Viscoplastic thin-walled columns response to pulse load. In: Dubina D (ed) Proc. ICTWS pp 415–430
134. Mania RJ (2011) Dynamic buckling of orthotropic viscoplastic column. Thin-Walled Struct 49(5):581–588
135. Mania RJ, Kowal-Michalska K (2007) Behaviour of composite columns of closed cross-section under in-plane compressive pulse loading. Thin-Walled Struct 45(10–11):125–129
136. Mania RJ, Kowal-Michalska K (2010) Elasto-plastic dynamic response of thin-walled columns subjected to pulse compression. In: Proc. of SSTA 2009, CRC Press pp 183–186
137. March H (1942) Buckling of flat plywood plates in compression, shear or combined compression and shear. Forest Products Laboratory Rep. 1316
138. Massey C (1971) The elastic buckling of orthotropic rectangular plates. Civ Eng Trans 13(1):63–65
139. Moellmann M, Goltermann P (1989) Interactive buckling in thin-walled beams, Part I: Theory, Part II: Applications. Int J Solid Struct 25(7):715–728
140. Moen CD, Schafer BW (2009) Elastic buckling of thin plates with holes in compression or bending. Thin-Walled Struct 47(12):1597–1607
141. Mulligan GP, Pekoz T (1984) Locally buckled thin-walled columns. J Struct Eng 110(11):2635–2654
142. Najafov AM, Sofiyev AH (2012) The non-linear dynamics of FGM truncated conical shells surrounded by an elastic medium. Int J Mech Sci 66:33–44
143. Noor AK (1975) Stability of multilayered composite plates. Fibre Sci Technol 8:81–89
144. Ovesy HR, Loughlan J, Ghannadpour SAM (2010) An investigation on the post-local-buckling analysis of I-section struts using finite strip method. Adv Steel Constr 6(1):662–677
145. Paczos P, Zielnica J (2008) Stability of orthotropic elastic-plastic open conical shells. Thin-Walled Struct 46(5):530–540
146. Papazouglou VJ, Tsouvalis NG (1995) Large deflection dynamic response of composite laminated plates under in-plane loads. Compos Struct 33:237–252
147. Petry D, Fahlbusch G (2000) Dynamic buckling of thin isotropic plates subjected to in-plane impact. Thin-Walled Struct 38:267–283
148. Pignataro M, Luongo A, Rizzi N (1985) On the effect of the local overall interaction on the post-buckling of uniformly compressed channel. Thin-Walled Struct 3:283–321
149. Pignataro M, Luongo A (1987) Asymmetric interactive buckling of thin-walled columns with initial imperfection. Thin-Walled Struct 3:365–385

150. Pignataro M, Luongo A (1987) Multiple Interactive Buckling of Thin-Walled Members in Compression. In: Proceedings of the International Colloqium on Stability of Plate and Shell Structures, University Ghent pp 235–240
151. Prahakara MK, Chia CY (1973) Postbuckling behaviour of rectangular orthotropic plates. J Mech Eng Sci 17:25–33
152. Reissner E, Stavsky Y (1961) Bending and stretching of certain types of heterogeneous aelotropic elastic plates. Trans ASME J Appl. Mech. 9:402–408
153. Rhodes J (2007) Some recent research on thin-walled members at the University of Strathclyde. Thin-Walled Struct 45(10–11):821–826
154. Rivallant S, Ferrero JF, Barrau JJ (2006) Dynamic buckling of foam stabilised composite skin. Compos Struct 72:486–493
155. Rossi B, Jaspart JP, Rasmussen KJR (2010) Combined distortional and overall flexural-torsional buckling of cold-formed stainless steel sections: Experimental investigations. J Struct Eng 136(4):354–360
156. Rzeszut K, Garstecki A (2009) Modeling of initial geometrical imperfections in stability analysis of thin-walled structures. J Theor Appl Mech 47(3):667–684
157. Rzeszut K, Garstecki A (2011) Thin-walled structures with slotted connections stability problems. Thin-Walled Struct 49(5):674–681
158. Seif M, Schafer BW (2010) Local buckling of structural steel shapes. J Constr Steel Res 66(10):1232–1247
159. Sevin E (1960) On the elastic bending of columns due to dynamic axial forces including effects of axial inertia. J Appl Mech Trans ASME pp 125–131
160. Seydel E (1933) The critical shear load of rectangular plate. NACA TM 705
161. Schafer BW, Adany, S (2006) Buckling analysis of cold-formed steel members using CUFSM: conventional and constrained finite strip methods. In: Proc. 18th International Specialty Conference on Cold-Formed Steel Structures, Orlando, Florida
162. Schokker A, Sridharan S, Kasagi A (1996) Dynamic buckling of composite shells. Comput Struct 59(1):43–55
163. Shariyat M (2009) Vibration and dynamic buckling control of imperfect hybrid FGM plates with temperature-dependent material properties subjected to thermo-electro-mechanical loading conditions. Compos Struct 88(2):240–252
164. Shariyat M (2011) A nonlinear double-superposition global-local theory for dynamic buckling of imperfect viscoelastic composite/sandwich plates: A hierarchical constitutive model. Compos Struct 93(7):1890–1899
165. Shariyat M (2011) A double-superposition global-local theory for vibration and dynamic buckling analyses of viscoelastic composite/sandwich plates: A complex modulus approach. Arch Appl Mech 81(9):1253–1268
166. Silva NMF, Silvestre N, Camotim D (2010) GBT formulation to analyse the buckling behaviour of FRP composite open-section thin-walled columns. Compos Struct 93(1):79–92
167. Simitses GJ (1987) Instability of dynamically loaded structures. Appl Mech Rev 40(10):1403–1408
168. Simitses GJ (1990) Dynamic stability of suddenly loaded structures. Springer Verlag, New York
169. Smith RCT (1944) The buckling of flat plywood plate in compression. Australian Council for Aeonautics Rep. 12
170. Sofiyev AH (2005) The stability of compositionally graded ceramic–metal cylindrical shells under aperiodic axial impulsive loading. Compos Struct 69:247–257
171. Sofiyev AH (2007) The buckling of functionally graded truncated conical shells under dynamic axial loading. J Sound Vib 305(4–5):808–826
172. Sofiyev AH (2010) Dynamic response of an FGM cylindrical shell under moving loads. Compos Struct 93(1):58–66
173. Sridharan S, Benito R (1984) Columns static and dynamic interactive buckling. J Eng Mech ASCE 110(1):49–65

174. Sridharan S, Ali MA (1986) Interactive buckling in thin-walled beam-columns. J Eng Mech ASCE 111(12):1470–1486
175. Sridharan S, Ali MA (1986) An improved interactive buckling analysis of thin-walled columns having doubly symmetric sections. Int J Solid Struct 22(4):429–443
176. Szymczak C (2011) Stability and initial post-buckling behaviour of frame system with vertical bracings. Thin-Walled Struct 49(5):669–673
177. Szyniszewski S, Smith BH, Hajjar JF, Arwade SR, Schafer BW (2012) Local buckling strength of steel foam sandwich panels. Thin-Walled Struct 59:11–19
178. Taub J (1934) Impact buckling of thin bars in the elastic range for any end condition. NACA TM 749
179. Teter A (2007) Static and dynamic interactive buckling of isotropic thin-walled closed columns with variable thickness. Thin-Walled Struct 45(10–11):936–940
180. Teter A (2010) Dynamic, multimode buckling of thin-walled columns subjected to in-plane pulse loading. Int J Non-Linear Mech 45(3):207–218
181. Teter A (2011) Dynamic critical load based on different stability criteria for coupled buckling of columns with stiffened open cross-sections. Thin-Walled Struct 49:589–595
182. Teter A, Kolakowski Z (2004) Interactive buckling and load carrying capacity of thin-walled beam-columns with intermediate stiffeners. Thin-Walled Struct 42:211–254
183. Teter A, Kolakowski Z (2005) Buckling of thin-walled composite structures with intermediate stiffeners. Compos Struct 69(4):421–428
184. Teter A, Kolakowski Z (2013) Coupled dynamic buckling of thin-walled composite columns with open cross-sections. Compos Struct 95:28–34
185. Thielemann W (1950) Contributions to the problem of buckling of orthotropic plates with special reference to plywood NACA TM 1263
186. Timoshenko SP, Gere JM (1961) Theory of elastic stability. McGraw-Hill
187. Ting TK, Surdenans J (1987) Buckling of rectangular orthotropic plates under biaxial loading. J Compos Mater 21:124–128
188. Tomski L, Szmidla J, Uzny S (2007) The local and global instability and vibration of systems subjected to non-conservative loading. Thin-Walled Struct 45(10–11):945–949
189. Tomski L, Uzny S (2008) Free vibrations and the stability of a geometrically non-linear column loaded by a follower force directed towards the positive pole. Int J Solids Struct 45(1):87–112
190. Tomski L, Podgorska-Brzdekiewicz I (2011) Global buckling and the interaction of initial imperfections of columns subjected to conservative loading. Thin-Walled Struct 49(5):596–603
191. Ungureanu V, Kotelko M, Mania RJ, Dubina D (2010) Plastic mechanisms database for thin-walled cold-formed steel members in compression and bending. Thin-Walled Struct 48(10–11):818–826
192. Vinson JR, Chou TW (1975) Composite Materials and their use in structures. Applied Science Publisher Ltd
193. Virella JC, Godoy LA, Suarez LE (2006) Dynamic buckling of anchored steel tanks subjected to horizontal earthquake excitation. J Constr Steel Res 62:521–531
194. Volmir SA (1967) Ustoiczivost deformirujemych system. Science, Moscow
195. Volmir SA (1972) Nieliniejnaja dinamika plastinok i oboloczek. Science, Moscow
196. Witmer EA, Pian TH (1962) Dynamic deformation and buckling of spherical shells under blast and impact loading. NASA TN D-1510
197. Weller T, Abramovich H, Yaffe R (1989) Dynamic buckling of beams and plates subjected to axial impact. Comput Struct 37:835–851
198. Xu X, Ma J, Lim CW, Chu H (2009) Dynamic local and global buckling of cylindrical shells under axial impact. Eng Struct 31(5):1132–1140
199. Xu X, Ma J, Lim CW, Zhang G (2010) Dynamic torsional buckling of cylindrical shells. Comput Struct 88(5–6):322–330
200. Yaffe R, Abramovich H (2003) Dynamic buckling of cylindrical stringer stiffened shells. Comput Struct 81:1031–1039

201. Yao Z, Rasmussen KJR (2012) Inelastic local buckling behaviour of perforated plates and sections under compression. Thin-Walled Struct 61:49–70
202. Yu C, Schafer BW (2007) Simulation of cold-formed steel beams in local and distortional buckling with applications to the direct strength method. J Constr Steel Res 63(5):581–590
203. Zhang T, Liu T, Zhao Y (2004) Nonlinear dynamic buckling of stiffened plates under in-plane impact load. J Zheijang Univ Sci 5(5):609–617
204. Zhang Z, Taheri F (2004) Dynamic pulse-buckling behavior of 'quasi-ductile' carbon/epoxy and E-glass/epoxy laminated composite beams. Compos Struct 64:269–274
205. Zielnica, J (2012) Buckling and stability of elastic-plastic sandwich conical shells. Steel Compos Struct 13(2):157–169
206. SAS IP, Inc (2007) ANSYS 11.1 html online documentation, SAS IP, Inc, 2007

Chapter 2
Theory of Thin Plates for Laminates

Thin isotropic, orthotropic or laminate plates with constant or widthwise variable material properties are considered in this monograph. Thin-walled beam-columns or girders composed of the above-mentioned plates are also analysed. In order to take into account all buckling modes (global, local and their interaction), the plate two-dimensional theory has been adopted to model the structures under analysis.

2.1 Basic Assumptions

Basic assumptions for thin plates were given by Kirchhoff for the linear classical thin plate theory (CPT) and by von Kármán and Marquerre for the nonlinear CPT. They made their assumptions for isotropic materials. Numerous authors have extended those assumptions for orthotropic or even for composite multilayer thin plates [1, 5]. The assumptions are as follows:

- the plate is homogeneous (for example, orthotropic homogenisation is made for a fibre composite—resin matrix and fibre-reinforcement);
- the plate is thin—other dimensions (length and width) are at least 10 times higher than the plate thickness;
- the material of the plate is deformable and it is subjected to Hooke's law;
- the plane stress state is considered for the plate—the stress acting in the plate plane dominates the plate behaviour, stresses acting in the direction normal to the plate plane are assumed to be zero;
- all strains (normal and shear) in the plate plane are low compared to unity and they are linear;
- the strains normal to the plate mid-surface are neglected (the plate thickness does not change after deformation)—this assumption is made according to Kirchhoff-Love;

T. Kubiak, *Static and Dynamic Buckling of Thin-Walled Plate Structures*,
DOI: 10.1007/978-3-319-00654-3_2,

- the straight lines normal to the mid-surface of the plate remain straight and normal to the mid-surface after deformation;
- there are no interactions in the normal direction between layers parallel to the middle surface;
- deflections of the plate can be considered in terms of nonlinear geometrical relations.

Additionally, it is assumed that the principal axes of orthotropy do not need to be parallel to the edges of analysed structures (a plate, a beam, column, a beam-column or a girder).

2.2 Geometrical Equations for Thin Plates

A two dimensional model of the plate has been assumed for thin plates and thin-walled beam-columns or girders. For a simpler description, a single plate (Fig. 2.1a) or each *i*-th strip (Fig. 2.1b) of the plate (or a wall of the girder) or each *i*-th wall of the girder (Fig. 2.1c) is referred to as a plate [3, 4].

To describe the middle surface strains for each plate, a complete strain tensor, i.e., with all nonlinear terms, has been assumed [1, 2]:

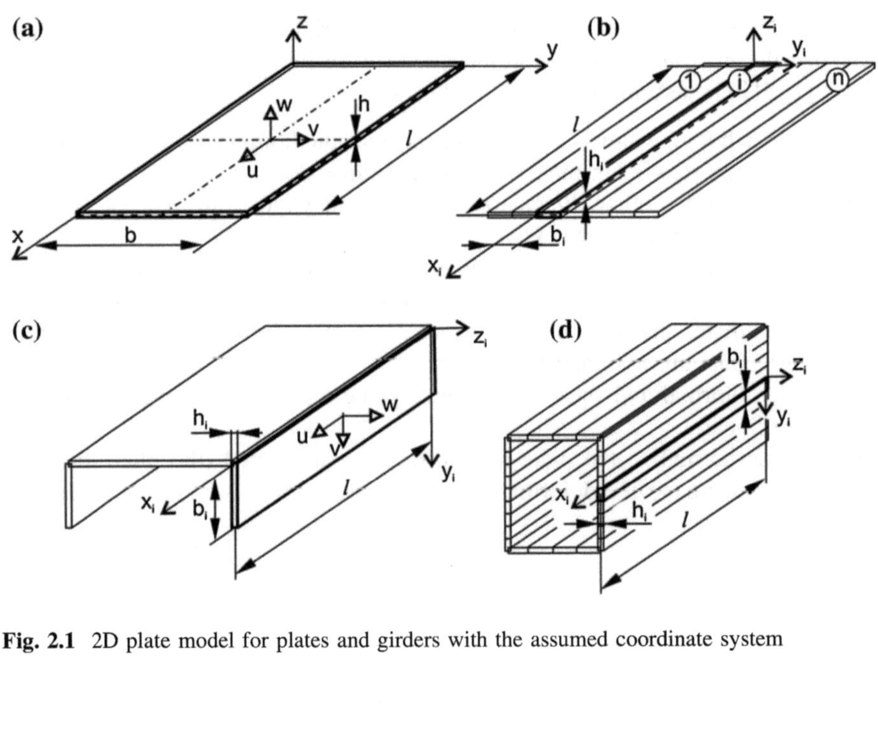

Fig. 2.1 2D plate model for plates and girders with the assumed coordinate system

$$
\begin{aligned}
\varepsilon_{ix}^{m} &= \frac{\partial u_i}{\partial x_i} + \frac{1}{2}\left(\frac{\partial w_i}{\partial x_i}\right)^2 + \frac{1}{2}\left(\frac{\partial u_i}{\partial x_i}\right)^2 + \frac{1}{2}\left(\frac{\partial v_i}{\partial x_i}\right)^2, \\
\varepsilon_{iy}^{m} &= \frac{\partial v_i}{\partial y_i} + \frac{1}{2}\left(\frac{\partial w_i}{\partial y_i}\right)^2 + \frac{1}{2}\left(\frac{\partial u_i}{\partial y_i}\right)^2 + \frac{1}{2}\left(\frac{\partial v_i}{\partial y_i}\right)^2, \\
2\varepsilon_{ixy}^{m} &= \gamma_{ixy}^{m} = \frac{\partial u_i}{\partial y_i} + \frac{\partial v_i}{\partial x_i} + \frac{\partial w_i}{\partial x_i}\frac{\partial w_i}{\partial y_i} + \frac{\partial u_i}{\partial x_i}\frac{\partial u_i}{\partial y_i} + \frac{\partial v_i}{\partial x_i}\frac{\partial v_i}{\partial y_i},
\end{aligned}
\tag{2.1}
$$

or in a shorter form:

$$
\begin{aligned}
\varepsilon_{ix}^{m} &= u_{i,x} + \frac{1}{2}(w_{i,x}^2 + u_{i,x}^2 + v_{i,x}^2), \\
\varepsilon_{iy}^{m} &= v_{i,y} + \frac{1}{2}(w_{i,y}^2 + u_{i,y}^2 + v_{i,y}^2), \\
\gamma_{ixy}^{m} &= u_{i,y} + v_{i,x} + w_{i,x}w_{i,y} + u_{i,x}u_{i,y} + v_{i,x}v_{i,y},
\end{aligned}
\tag{2.2}
$$

where: u_i, v_i, w_i are displacements parallel to the respective axes x_i, y_i, z_i of the local Cartesian system of co-ordinates, whose plane x_iy_i coincides with the middle surface of the *i*-th plate before its buckling (Fig. 2.1).

In the majority of publications devoted to the structure stability, the terms$(u_{i,x}^2 + v_{i,x}^2)$, $(u_{i,y}^2 + v_{i,y}^2)$ and $(u_{i,x}u_{i,y} + v_{i,x}v_{i,y})$, i.e., the strain tensor components in (2.2), in general are neglected for ε_{ix}, ε_{iy}, γ_{ixy}, respectively.

The changes in the bending and twisting curvatures of the middle surface are assumed according to [6, 7] as follows:

$$
\begin{aligned}
\kappa_{ix} &= -\frac{\partial^2 w_i}{\partial x_i^2} = -w_{i,xx}, \\
\kappa_{iy} &= -\frac{\partial^2 w_i}{\partial y_i^2} = -w_{i,yy}, \\
\kappa_{ixy} &= -\frac{\partial^2 w_i}{\partial x_i \partial y_i} = -w_{i,xy}.
\end{aligned}
\tag{2.3}
$$

The geometrical relationships given by Eqs. (2.2) and (2.3) allow one to consider both out-of-plane and in-plane bending of the plate.

For the laminated plate (Fig. 2.2), where there is a *p* number of plies, the strains of the *k*-th ply can be related to the strains and the curvatures of the middle surface of the laminate at $z = 0$ in the form [1]:

$$
\{\bar{\varepsilon}\} = \begin{Bmatrix} \varepsilon_{ix} \\ \varepsilon_{iy} \\ \gamma_{ixy} \end{Bmatrix} = \begin{Bmatrix} \varepsilon_{ix}^{m} \\ \varepsilon_{iy}^{m} \\ \gamma_{ixy}^{m} \end{Bmatrix} + z \begin{Bmatrix} \kappa_{ix} \\ \kappa_{iy} \\ 2\kappa_{ixy} \end{Bmatrix},
\tag{2.4}
$$

where $z_{k-1} \le z \le z_k$ (Fig. 2.2) and for $k = p = 1$, $z0 = -h/2$, $z_1 = h/2$.

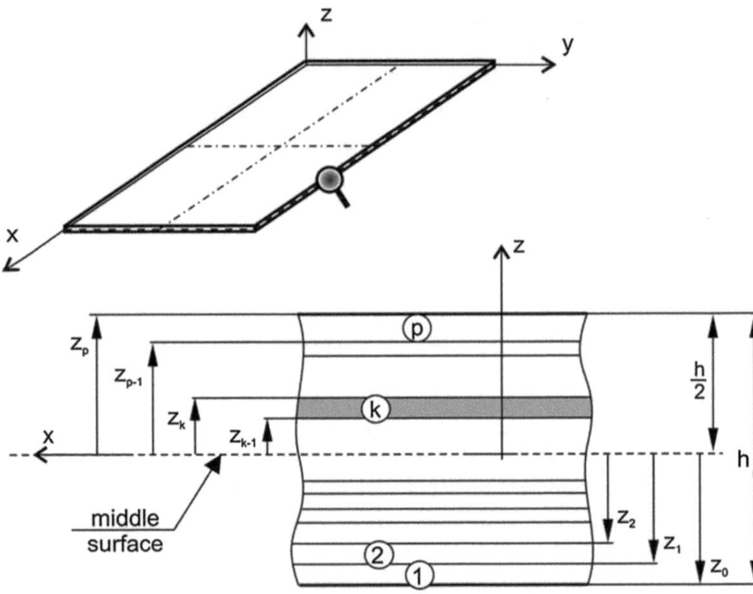

Fig. 2.2 Assumed coordinate system for the layered plate

2.3 Constitutive Equations for Laminates

Let us consider one rectangular ply of the laminate with principal axes of orthotropy *1* and *2* parallel to ply edges (Fig. 2.3).

Similarly as in the previous paragraph, let us consider an *i*-th plate or strip of the structure under analysis. The stress-strain relationship for such a plate is the same as for an orthotropic plate and can be written in the following form [1]:

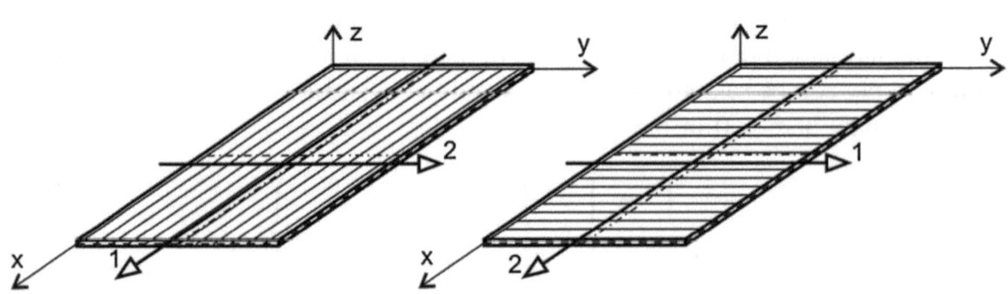

Fig. 2.3 Principal axes of orthotropy for lamina

$$\begin{aligned}\sigma_{i1} &= \frac{E_{i1}}{1-\nu_{i12}\nu_{i21}}(\varepsilon_{i1}+\nu_{i21}\varepsilon_{i2}),\\ \sigma_{i2} &= \frac{E_{i2}}{1-\nu_{i12}\nu_{i21}}(\nu_{i12}\varepsilon_{i1}+\varepsilon_{i2}),\\ \tau_{i12} &= G_{i12}\gamma_{i12} = 2G_{i12}\varepsilon_{i12},\end{aligned} \tag{2.5}$$

where E_{i1}, E_{i2} is the Young's modulus in longitudinal *1* and transverse *2* direction, correspondingly; ν_{i12} is the Poisson's ratio for which strains are in longitudinal direction *1* and stress in transverse direction *2*, G_{i12} is the shear modulus (Kirchhoff's modulus) in plane *12*.

In further equations in this section, the subscript *i* denoting an *i*-th plate or strip is omitted because all the equations presented correspond to one plate or strip only.

Equations (2.5) written in a matrix form are as follows:

$$\begin{Bmatrix}\sigma_1\\ \sigma_2\\ \tau_{12}\end{Bmatrix} = \begin{bmatrix}Q_{11} & Q_{12} & 0\\ Q_{21} & Q_{22} & 0\\ 0 & 0 & Q_{66}\end{bmatrix}\begin{Bmatrix}\varepsilon_1\\ \varepsilon_2\\ \gamma_{12}\end{Bmatrix}, \tag{2.6}$$

or in a more convenient form:

$$\{\sigma\} = [Q]\{\varepsilon\}, \tag{2.7}$$

where:

$$\begin{aligned}Q_{11} &= \frac{E_1}{1-\nu_{12}\nu_{21}},\\ Q_{12} &= Q_{21} = \nu_{21}\frac{E_1}{1-\nu_{12}\nu_{21}} = \nu_{12}\frac{E_2}{1-\nu_{12}\nu_{21}},\\ Q_{22} &= \frac{E_2}{1-\nu_{12}\nu_{21}},\\ Q_{66} &= G_{12}.\end{aligned} \tag{2.8}$$

The Young's modulus and Poisson's ratios occurring in (2.5) and (2.8) according to the Betty-Maxwell theorem or according to the symmetry condition of the stress tensor ($Q_{12} = Q_{21}$) should fulfil the following relation:

$$E_1\nu_{21} = E_2\nu_{12}. \tag{2.9}$$

Fibres in individual plies of laminates are arranged at different angles to the plate edges. It means that the principal axes of orthotropy are rotated at an angle θ in relation to the coordinate system adopted for the entire plate (Fig. 2.4).

For the plate presented in Fig. 2.4, stress-strain relationship (2.5) should be transformed from the local *1-2* coordinate system to the global *xy* one. The constitutive equations in the local coordinate system are given by (2.6), and in the global coordinates, they can be written as follows:

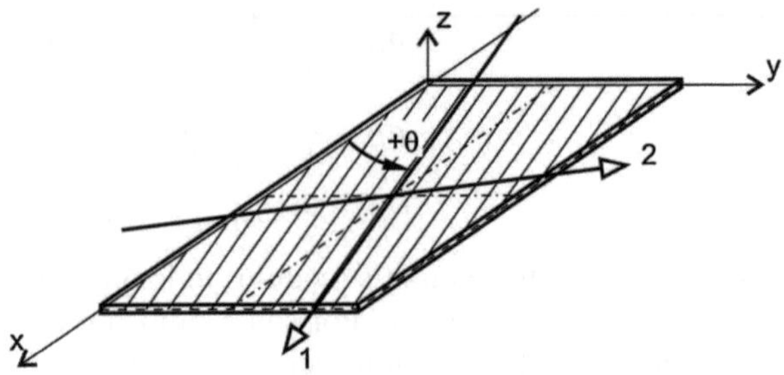

Fig. 2.4 Fibre orientation in the composite ply

$$\begin{Bmatrix} \sigma_x \\ \sigma_y \\ \tau_{xy} \end{Bmatrix} = \begin{bmatrix} \overline{Q}_{11} & \overline{Q}_{12} & \overline{Q}_{16} \\ \overline{Q}_{21} & \overline{Q}_{22} & \overline{Q}_{26} \\ \overline{Q}_{61} & \overline{Q}_{62} & \overline{Q}_{66} \end{bmatrix} \begin{Bmatrix} \varepsilon_x \\ \varepsilon_y \\ \gamma_{xy} \end{Bmatrix}, \tag{2.10}$$

or shorter:

$$\{\overline{\sigma}\} = \left[\overline{Q}\right]\{\overline{\varepsilon}\}, \tag{2.11}$$

where the elements of the elasticity matrix $\left[\overline{Q}\right]$are expressed by material properties (E_1, E_2 ν_{12} and G_{12}) and the angle of the declination θ between the global and local coordinate systems. The relation between the elasticity matrix in local [Q] and global $\left[\overline{Q}\right]$ coordinate systems can be derived taking into account the relation between stresses in both coordinate systems and strains in both the systems. The stress transformation equations are as follows:

$$\begin{aligned} \sigma_1 &= \sigma_x \cos^2\theta + \sigma_y \sin^2\theta + 2\tau_x \cos\theta \sin\theta, \\ \sigma_2 &= \sigma_x \sin^2\theta + \sigma_y \cos^2\theta - 2\tau_x \cos\theta \sin\theta, \\ \tau_{12} &= -\sigma_x \cos\theta \sin\theta + \sigma_y \cos\theta \sin\theta + \tau_x(\cos^2\theta - \sin^2\theta), \end{aligned} \tag{2.12}$$

or in the matrix form:

$$\begin{Bmatrix} \sigma_1 \\ \sigma_2 \\ \tau_{12} \end{Bmatrix} = \begin{bmatrix} c^2 & s^2 & 2cs \\ s^2 & c^2 & -2cs \\ -cs & cs & c^2 - s^2 \end{bmatrix} \begin{Bmatrix} \sigma_x \\ \sigma_y \\ \tau_{xy} \end{Bmatrix}, \tag{2.13}$$

or shorter:

$$\{\sigma\} = [T]\{\overline{\sigma}\}, \tag{2.14}$$

where: $\{\sigma\}$, $\{\overline{\sigma}\}$ are vectors of stresses in the local and global coordinate systems, correspondingly, $c = \cos\theta$, $s = \sin\theta$ and [T] is the transformation matrix. To find the stress in the global coordinate system having the stress in the coordinate system corresponding to the principle axes of orthotropy, the following relation should be used:

$$\{\overline{\sigma}\} = [T]^{-1}\{\sigma\}, \tag{2.15}$$

For strains, the transformation equations can be written as follows:

$$
\begin{aligned}
\varepsilon_1 &= \varepsilon_x \cos^2\theta + \varepsilon_y \sin^2\theta + 2\varepsilon_{xy}\cos\theta\sin\theta, \\
\varepsilon_2 &= \varepsilon_x \sin^2\theta + \varepsilon_y \cos^2\theta - 2\varepsilon_{xy}\cos\theta\sin\theta, \\
\varepsilon_{12} &= -\frac{\gamma_{12}}{2} = -\varepsilon_x\cos\theta\sin\theta + \varepsilon_y\cos\theta\sin\theta + \varepsilon_{xy}(\cos^2\theta - \sin^2\theta),
\end{aligned}
\tag{2.16}
$$

or in the matrix form:

$$
\begin{Bmatrix} \varepsilon_1 \\ \varepsilon_2 \\ \varepsilon_{12} \end{Bmatrix} = \begin{bmatrix} c^2 & s^2 & 2cs \\ s^2 & c^2 & -2cs \\ -cs & cs & c^2 - s^2 \end{bmatrix} \begin{Bmatrix} \varepsilon_x \\ \varepsilon_y \\ \varepsilon_{xy} \end{Bmatrix}, \tag{2.17}
$$

or

$$
\{\varepsilon\} = \begin{Bmatrix} \varepsilon_1 \\ \varepsilon_2 \\ \varepsilon_{12} \end{Bmatrix} = [T] \begin{Bmatrix} \varepsilon_x \\ \varepsilon_y \\ \varepsilon_{xy} \end{Bmatrix} = [T]\{\bar{\varepsilon}\}. \tag{2.18}
$$

From the theory of elasticity, it is well known that strain and stress transformations are made in the same way—the transformation matrix [T] is the same (see (2.13) and (2.17)). It should be noted that in (2.16)–(2.18), the shear strains ε_{12}, ε_{xy} in the strain vectors appear instead of the shear angles γ_{12}, γ_{xy}, which are used in constitutive equations (2.6) and (2.10). The relation between strain vectors including the shear strain or the shear angle can be written as follows:

$$
\begin{aligned}
\{\varepsilon\} &= \begin{Bmatrix} \varepsilon_1 \\ \varepsilon_2 \\ \gamma_{12} \end{Bmatrix} = \begin{bmatrix} 1 & 0 & 0 \\ 0 & 1 & 0 \\ 0 & 0 & 2 \end{bmatrix} \begin{Bmatrix} \varepsilon_1 \\ \varepsilon_2 \\ \varepsilon_{12} \end{Bmatrix} = [R] \begin{Bmatrix} \varepsilon_1 \\ \varepsilon_2 \\ \varepsilon_{12} \end{Bmatrix}, \\
\{\bar{\varepsilon}\} &= \begin{Bmatrix} \varepsilon_x \\ \varepsilon_y \\ \gamma_{xy} \end{Bmatrix} = \begin{bmatrix} 1 & 0 & 0 \\ 0 & 1 & 0 \\ 0 & 0 & 2 \end{bmatrix} \begin{Bmatrix} \varepsilon_x \\ \varepsilon_y \\ \varepsilon_{xy} \end{Bmatrix} = [R] \begin{Bmatrix} \varepsilon_x \\ \varepsilon_y \\ \varepsilon_{xy} \end{Bmatrix}.
\end{aligned}
\tag{2.19}
$$

Now, using (2.18) and (2.19), the strain transformation can be written as follows:

$$
\{\varepsilon\} = [R] \begin{Bmatrix} \varepsilon_1 \\ \varepsilon_2 \\ \varepsilon_{12} \end{Bmatrix} = [R][T] \begin{Bmatrix} \varepsilon_x \\ \varepsilon_y \\ \varepsilon_{xy} \end{Bmatrix} = [R][T][R]^{-1}\{\bar{\varepsilon}\}. \tag{2.20}
$$

However, $[R][T][R]^{-1}$ can be shown to be $[T]^{-T}$, then (2.20) has the form:

$$
\{\varepsilon\} = [T]^{-T}\{\bar{\varepsilon}\}. \tag{2.21}
$$

Substituting the constitutive equations in local coordinate system (2.7) into stress transformation equations (2.15), the following is obtained:

$$\{\overline{\sigma}\} = [T]^{-1}[Q]\{\varepsilon\}. \tag{2.22}$$

After substituting the strain transformation, i.e., (2.21) into (2.22), we obtain the following relation:

$$\{\overline{\sigma}\} = [T]^{-1}[Q][T]^{-T}\{\overline{\varepsilon}\}, \tag{2.23}$$

which is a constitutive equation in the global coordinate system. Comparing (2.23) and (2.11), one obtains the elasticity matrix transformation:

$$[\overline{Q}] = [T]^{-1}[Q][T]^{-T}, \tag{2.24}$$

or in the full form:

$$\begin{bmatrix} \overline{Q}_{11} & \overline{Q}_{12} & \overline{Q}_{16} \\ \overline{Q}_{21} & \overline{Q}_{22} & \overline{Q}_{26} \\ \overline{Q}_{61} & \overline{Q}_{62} & \overline{Q}_{66} \end{bmatrix} = \begin{bmatrix} c^2 & s^2 & -2cs \\ s^2 & c^2 & 2cs \\ cs & -cs & c^2-s^2 \end{bmatrix} \begin{bmatrix} Q_{11} & Q_{12} & 0 \\ Q_{21} & Q_{22} & 0 \\ 0 & 0 & Q_{66} \end{bmatrix} \begin{bmatrix} c^2 & s^2 & cs \\ s^2 & c^2 & -cs \\ -2cs & 2cs & c^2-s^2 \end{bmatrix}. \tag{2.25}$$

Taking into account (2.25), all elements of the elasticity matrix $[\bar{Q}]$ in the global coordinate system can be calculated and they are as follows:

$$\begin{aligned} \overline{Q}_{11} &= Q_{11}c^4 + 2(Q_{12} + 2Q_{66})c^2s^2 + Q_{22}s^4, \\ \overline{Q}_{12} &= \overline{Q}_{21} = (Q_{11} + Q_{22} - 4Q_{66})c^2s^2 + Q_{12}(c^4 + s^4), \\ \overline{Q}_{22} &= Q_{11}s^4 + 2(Q_{12} + 2Q_{66})c^2s^2 + Q_{22}c^4, \\ \overline{Q}_{16} &= \overline{Q}_{61} = (Q_{11} - Q_{12} - 2Q_{66})c^3s + (Q_{12} - Q_{22} + 2Q_{66})cs^3, \\ \overline{Q}_{26} &= \overline{Q}_{62} = (Q_{11} - Q_{12} - 2Q_{66})cs^3 + (Q_{12} - Q_{22} + 2Q_{66})c^3s, \\ \overline{Q}_{66} &= (Q_{11} + Q_{22} - 2Q_{12} - 2Q_{66})c^2s^2 + Q_{66}(c^4 + s^4). \end{aligned} \tag{2.26}$$

Summing the above, the constitutive equations can be rewritten as follows:

- for the k-th ply of laminates:

$$\begin{aligned} \begin{Bmatrix} \sigma_x \\ \sigma_y \\ \tau_{xy} \end{Bmatrix}_k &= \begin{bmatrix} \overline{Q}_{11} & \overline{Q}_{12} & \overline{Q}_{16} \\ \overline{Q}_{21} & \overline{Q}_{22} & \overline{Q}_{26} \\ \overline{Q}_{61} & \overline{Q}_{62} & \overline{Q}_{66} \end{bmatrix}_k \begin{Bmatrix} \varepsilon_x \\ \varepsilon_y \\ \gamma_{xy} \end{Bmatrix}_k \\ &= \begin{bmatrix} \overline{Q}_{11} & \overline{Q}_{12} & \overline{Q}_{16} \\ \overline{Q}_{21} & \overline{Q}_{22} & \overline{Q}_{26} \\ \overline{Q}_{61} & \overline{Q}_{62} & \overline{Q}_{66} \end{bmatrix}_k \left\{ \begin{Bmatrix} \varepsilon_x^m \\ \varepsilon_y^m \\ \gamma_{xy}^m \end{Bmatrix} + z \begin{Bmatrix} \kappa_x \\ \kappa_y \\ 2\kappa_{xy} \end{Bmatrix} \right\}, \end{aligned} \tag{2.27}$$

It should be noted that for each ply, the elasticity matrix $[\overline{Q}]_k$ can be different, then the stress can vary through the thickness of the laminate, not necessarily linearly, as it is in the case of strain (if all plies are in the elastic range).

- for the orthotropic plate with the principal axes of orthotropy parallel to the plate (strip) edges:

$$\begin{Bmatrix} \sigma_x \\ \sigma_y \\ \tau_{xy} \end{Bmatrix} = \begin{bmatrix} \frac{E_x}{1-\nu_{xy}\nu_{yx}} & \nu_{yx}\frac{E_x}{1-\nu_{xy}\nu_{yx}} & 0 \\ \nu_{xy}\frac{E_y}{1-\nu_{xy}\nu_{yx}} & \frac{E_y}{1-\nu_{xy}\nu_{yx}} & 0 \\ 0 & 0 & G_{xy} \end{bmatrix} \begin{Bmatrix} \varepsilon_x \\ \varepsilon_y \\ \gamma_{xy} \end{Bmatrix}, \tag{2.28}$$

- for the isotropic plate (wall of beam-columns):

$$\begin{Bmatrix} \sigma_x \\ \sigma_y \\ \tau_{xy} \end{Bmatrix} = \frac{E}{1-\nu^2} \begin{bmatrix} 1 & \nu & 0 \\ \nu & 1 & 0 \\ 0 & 0 & \frac{1-\nu}{2} \end{bmatrix} \begin{Bmatrix} \varepsilon_x \\ \varepsilon_y \\ \gamma_{xy} \end{Bmatrix}. \tag{2.29}$$

2.4 Generalized Sectional Forces

By generalized sectional forces are meant here sectional forces and moments (Fig. 2.5) dependent on the stress in the section under consideration.

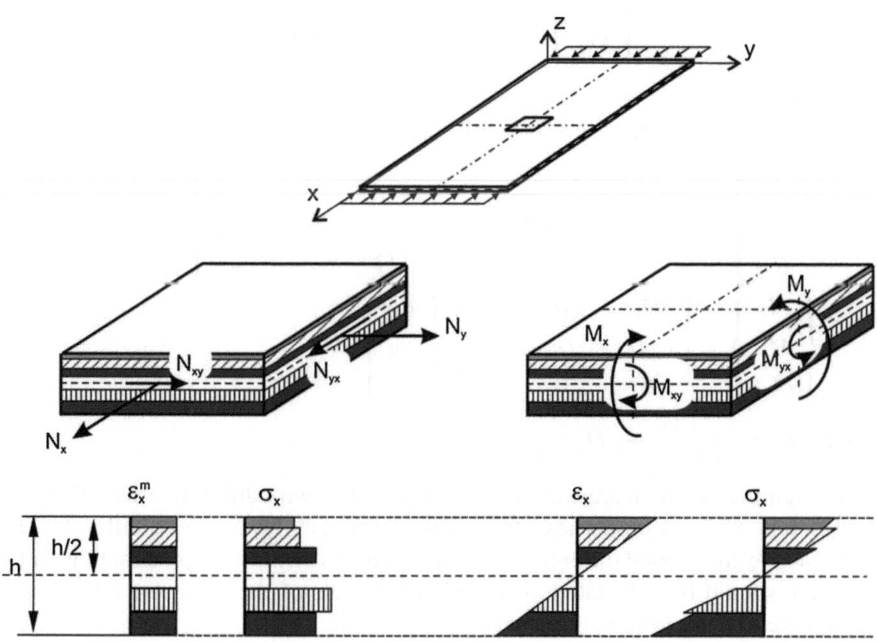

Fig. 2.5 Sectional forces and strains and a stress distribution in laminates

In the case of an isotropic plate or an orthotropic plate with the principal axes of orthotropy parallel to plate edges, the resultant moments and forces can be calculated as follows:

$$\begin{Bmatrix} N_x \\ N_y \\ N_{xy} \end{Bmatrix} = \{N_i\} = \int_{-\frac{h_i}{2}}^{\frac{h_i}{2}} \begin{Bmatrix} \sigma_x \\ \sigma_y \\ \tau_{xy} \end{Bmatrix} dz,$$
$$\begin{Bmatrix} M_x \\ M_y \\ M_{xy} \end{Bmatrix} = \{M_i\} = \int_{-\frac{h_i}{2}}^{\frac{h_i}{2}} \begin{Bmatrix} \sigma_x \\ \sigma_y \\ \tau_{xy} \end{Bmatrix} z dz, \tag{2.30}$$

where $N_{\mathrm{xy}} = N_{\mathrm{yx}}$ and $M_{\mathrm{xy}} = M_{\mathrm{yx}}$.

Substituting the stress-strain relations from the previous sections (Sect. 2.3), the sectional moments and forces:

- for the i-th isotropic strip or wall of the beam-column are expressed by:

$$\begin{Bmatrix} N_x \\ N_y \\ N_{xy} \end{Bmatrix} = \frac{Eh}{1-\nu^2} \begin{bmatrix} 1 & \nu & 0 \\ \nu & 1 & 0 \\ 0 & 0 & \frac{1-\nu}{2} \end{bmatrix} \begin{Bmatrix} \varepsilon_x^m \\ \varepsilon_y^m \\ \gamma_{xy}^m \end{Bmatrix},$$
$$\begin{Bmatrix} M_x \\ M_y \\ M_{xy} \end{Bmatrix} = D \begin{bmatrix} 1 & \nu & 0 \\ \nu & 1 & 0 \\ 0 & 0 & 1-\nu \end{bmatrix} \begin{Bmatrix} \kappa_x \\ \kappa_y \\ \kappa_{xy} \end{Bmatrix}, \tag{2.31}$$

where: $D = \frac{Eh^3}{12(1-\nu^2)}$

- for the i-th orthotropic strip or wall, they are:

$$\begin{Bmatrix} N_x \\ N_y \\ N_{xy} \end{Bmatrix} = \frac{h}{1-\nu_{xy}\nu_{yx}} \begin{bmatrix} E_x & \nu_{yx}E_x & 0 \\ \nu_{xy}E_y & E_y & 0 \\ 0 & 0 & (1-\nu_{xy}\nu_{yx})G_{xy} \end{bmatrix} \begin{Bmatrix} \varepsilon_x^m \\ \varepsilon_y^m \\ \gamma_{xy}^m \end{Bmatrix},$$
$$\begin{Bmatrix} M_x \\ M_y \\ M_{xy} \end{Bmatrix} = \begin{bmatrix} D_x & \nu_{yx}D_x & 0 \\ \nu_{xy}D_y & D_y & 0 \\ 0 & 0 & D_{xy} \end{bmatrix} \begin{Bmatrix} \kappa_x \\ \kappa_y \\ \kappa_{xy} \end{Bmatrix}, \tag{2.32}$$

where: $D_x = \frac{E_x h^3}{12(1-\nu_{xy}\nu_{yx})}$, $D_y = \frac{E_y h^3}{12(1-\nu_{xy}\nu_{yx})}$, $D_{xy} = \frac{G_{xy}h^3}{6}$.

For laminates which are composed of many plies with different orientation and/or different material properties, stress tensors for each layer can be different. Due to the above-mentioned differences, the resultant moments and forces acting on the laminate should be calculated as a sum of integrals of the stress for all laminate plies, taken in the following manner:

$$\begin{Bmatrix} N_x \\ N_y \\ N_{xy} \end{Bmatrix} = \sum_{k=1}^{p} \int_{z_{k-1}}^{z_k} \begin{Bmatrix} \sigma_x \\ \sigma_y \\ \tau_{xy} \end{Bmatrix}_k dz$$
$$= \sum_{k=1}^{p} \begin{bmatrix} \overline{Q}_{11} & \overline{Q}_{12} & \overline{Q}_{16} \\ \overline{Q}_{21} & \overline{Q}_{22} & \overline{Q}_{26} \\ \overline{Q}_{61} & \overline{Q}_{62} & \overline{Q}_{66} \end{bmatrix}_{ik} \left[\int_{z_{k-1}}^{z_k} \begin{Bmatrix} \varepsilon_x^m \\ \varepsilon_y^m \\ \gamma_{xy}^m \end{Bmatrix} dz + \int_{z_{k-1}}^{z_k} \begin{Bmatrix} \kappa_x \\ \kappa_y \\ 2\kappa_{xy} \end{Bmatrix} zdz \right], \tag{2.33}$$

and

$$\begin{Bmatrix} M_x \\ M_y \\ M_{xy} \end{Bmatrix} = \sum_{k=1}^{p} \int_{z_{k-1}}^{z_k} \begin{Bmatrix} \sigma_x \\ \sigma_y \\ \tau_{xy} \end{Bmatrix}_k zdz$$
$$= \sum_{k=1}^{p} \begin{bmatrix} \overline{Q}_{11} & \overline{Q}_{12} & \overline{Q}_{16} \\ \overline{Q}_{21} & \overline{Q}_{22} & \overline{Q}_{26} \\ \overline{Q}_{61} & \overline{Q}_{62} & \overline{Q}_{66} \end{bmatrix}_{ik} \left[\int_{z_{k-1}}^{z_k} \begin{Bmatrix} \varepsilon_x^m \\ \varepsilon_y^m \\ \gamma_{xy}^m \end{Bmatrix} zdz + \int_{z_{k-1}}^{z_k} \begin{Bmatrix} \kappa_x \\ \kappa_y \\ 2\kappa_{xy} \end{Bmatrix} z^2dz \right]. \tag{2.34}$$

In the above equations, the stiffness matrices $[\overline{Q}]_k$ are outside the integral over each layer because their elements are constant across the thickness of every particular layer.

As we know, all strains ε and all curvatures κ are not functions of z, but they refer to the middle surface, so they can be drawn outside the summation signs. Thus, (2.33) and (2.34) can be written as:

$$\begin{Bmatrix} N_x \\ N_y \\ N_{xy} \\ M_x \\ M_y \\ M_{xy} \end{Bmatrix} = \begin{bmatrix} A_{11} & A_{12} & A_{16} & B_{11} & B_{12} & B_{16} \\ A_{21} & A_{22} & A_{26} & B_{21} & B_{22} & B_{26} \\ A_{61} & A_{62} & A_{66} & B_{61} & B_{62} & B_{66} \\ B_{11} & B_{12} & B_{16} & D_{11} & D_{12} & D_{16} \\ B_{21} & B_{22} & B_{26} & D_{21} & D_{22} & D_{26} \\ B_{61} & B_{62} & B_{66} & D_{61} & D_{62} & D_{66} \end{bmatrix}_i \begin{Bmatrix} \varepsilon_x^m \\ \varepsilon_y^m \\ \gamma_{xy}^m \\ \kappa_x \\ \kappa_y \\ \kappa_{xy} \end{Bmatrix}, \tag{2.35}$$

or in a more convenient form:

$$\begin{Bmatrix} \{N\} \\ \{M\} \end{Bmatrix} = \begin{bmatrix} [A] & [B] \\ [B] & [D] \end{bmatrix} \begin{Bmatrix} \{\varepsilon^m\} \\ \{\kappa\} \end{Bmatrix}, \tag{2.36}$$

where:

$$\begin{aligned} A_{pq} &= \sum_{k=1}^{n} \left(\overline{Q}_{pq}\right)_k (z_k - z_{k-1}), \\ B_{pq} &= \frac{1}{2} \sum_{k=1}^{n} \left(\overline{Q}_{pq}\right)_k \left(z_k^2 - z_{k-1}^2\right), \\ D_{pq} &= \frac{1}{3} \sum_{k=1}^{n} \left(\overline{Q}_{pq}\right)_k \left(z_k^3 - z_{k-1}^3\right), \end{aligned} \tag{2.37}$$

and

$$A_{pq} = A_{qp},\ B_{pq} = B_{qp},\ D_{pq} = D_{qp}. \tag{2.38}$$

In (2.36), the sub-matrix [A] is an extensional stiffness matrix, [D] is a bending stiffness matrix, and [B] is a bending-extension coupling the stiffness matrix. If all elements of the sub-matrix [B] are not equal to zero, then, for example, the deformation of the laminate subjected to tension load is not only extension but also bending and/or twisting (Fig. 2.6a). Another example for non-zero sub-matrix [B] elements is such that during bending the laminate is bent and also suffers from extension of the middle surface (Fig. 2.6b).

The laminates can represent a few special cases of layers alignment and these are:

- symmetrical structure—the laminate consists of an even number of layers arranged symmetrically about the middle surface. Thus, the following stiffness matrix elements are equal to zero:

$$B_{pq} = A_{16} = A_{61} = A_{26} = A_{62} = D_{16} = D_{61} = D_{26} = D_{62} = 0, \tag{2.39}$$

- regular symmetric cross-ply laminate—called a quasi-orthotropic material—the laminate consists of an odd number of layers arranged symmetrically with respect to the middle surface and the lamina pairs are oriented in such a way that the principal material direction corresponds to plate edges (plies are oriented at 0 or 90 degrees to the longitudinal direction of the considered plate). For such a laminate, the following elements of the stiffness matrix are equal to zero:

$$B_{pq} = A_{16} = A_{61} = A_{26} = A_{62} = D_{16} = D_{61} = D_{26} = D_{62} = 0, \tag{2.40}$$

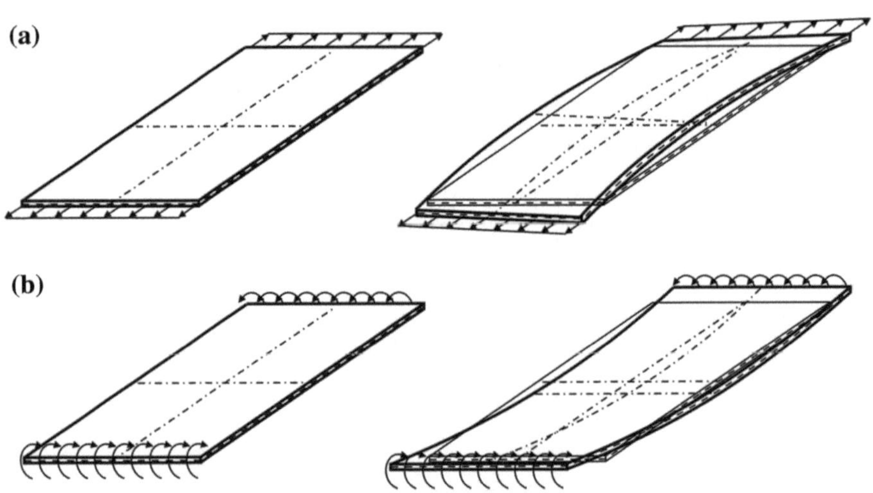

Fig. 2.6 Possible deflection of non-quasi-isotropic laminates

- regular symmetric angle-ply laminate—the laminate consists of an odd number of plies with equal thickness and principal material properties of each layers are arranged with opposite signs of the angle of orientation (for example, [+ θ/−θ/ + θ]). In this case, the following elements of the stiffness matrix diminish:

$$B_{pq} = 0, \tag{2.41}$$

- antisymmetric cross-ply laminate—the laminate consists of an even number of layers of the same thickness, laid on each other with the principal axes of orthotropy alternating at 0 and 90 degrees to the laminate axes. The known information about elements of the stiffness matrix is as follows:

$$\begin{aligned} &A_{16} = A_{61} = A_{26} = A_{62} = 0, \\ &D_{16} = D_{61} = D_{26} = D_{62} = 0, \\ &B_{12} = B_{21} = B_{16} = B_{61} = B_{26} = B_{62} = B_{66} = 0, \\ &B_{11} = -B_{22} \neq 0, \end{aligned} \tag{2.42}$$

- antisymmetric angle-ply laminate—the laminate consists of an even number of layers of the same thickness and its plies are arranged in pairs at an angle +θ and −θ, respectively. The elements of the stiffness matrix with the zero value are as follows:

$$\begin{aligned} &A_{16} = A_{61} = A_{26} = A_{62} = 0, \\ &D_{16} = D_{61} = D_{26} = D_{62} = 0, \\ &B_{11} = B_{12} = B_{21} = B_{22} = B_{66} = 0. \end{aligned} \tag{2.43}$$

The examples taken for calculation and presented in this study are obtained only for isotropic structures, orthotropic structures with the principal axes of orthotropy parallel to edges of structures and laminates with a symmetrical arrangement of the layers. For all above-mentioned cases, the expression describing sectional moments and forces can be written as:

$$\begin{aligned} \begin{Bmatrix} N_x \\ N_y \\ N_{xy} \end{Bmatrix} &= \begin{bmatrix} A_{11} & A_{12} & 0 \\ A_{21} & A_{22} & 0 \\ 0 & 0 & A_{66} \end{bmatrix} \begin{Bmatrix} \varepsilon_x^m \\ \varepsilon_y^m \\ \gamma_{xy}^m \end{Bmatrix}, \\ \begin{Bmatrix} M_x \\ M_y \\ M_{xy} \end{Bmatrix} &= \begin{bmatrix} D_{11} & D_{12} & 0 \\ D_{21} & D_{22} & 0 \\ 0 & 0 & D_{66} \end{bmatrix} \begin{Bmatrix} \kappa_x \\ \kappa_y \\ \kappa_{xy} \end{Bmatrix}, \end{aligned} \tag{2.44}$$

where all A_{pq} and D_{pq} should fulfil (2.38) and for:

- symmetric laminates, are described by (2.37),
- orthotropic plates (or strips), they are:

$$
\begin{aligned}
&A_{11}=\frac{E_x h}{1-\nu_{xy}\nu_{yx}}, && D_{11}=\frac{E_x h^3}{12\left(1-\nu_{xy}\nu_{yx}\right)},\\
&A_{12}=A_{21}=\frac{\nu_{yx}E_x h}{1-\nu_{xy}\nu_{yx}}=\frac{\nu_{xy}E_y h}{1-\nu_{xy}\nu_{yx}}, && D_{12}=D_{21}=\frac{\nu_{yx}E_x h^3}{12\left(1-\nu_{xy}\nu_{yx}\right)}=\frac{\nu_{xy}E_y h^3}{12\left(1-\nu_{xy}\nu_{yx}\right)},\\
&A_{22}=\frac{E_y h}{1-\nu_{xy}\nu_{yx}}, && D_{22}=\frac{E_y h^3}{12\left(1-\nu_{xy}\nu_{yx}\right)},\\
&A_{66}=G_{xy}h, && D_{66}=\frac{G_{xy}h^3}{6}.
\end{aligned}
\tag{2.45}
$$

- isotropic plates, they are as follows:

$$
\begin{aligned}
&A_{11}=A_{22}=\frac{Eh}{1-\nu^2}, && D_{11}=D_{22}=\frac{Eh^3}{12(1-\nu^2)},\\
&A_{12}=A_{21}=\frac{\nu Eh}{1-\nu^2}, && D_{12}=D_{21}=\frac{\nu Eh^3}{12(1-\nu^2)},\\
&A_{66}=Gh=\frac{Eh}{2(1+\nu)}, && D_{66}=\frac{Gh^3}{6}=\frac{Eh^3}{12(1+\nu)}.
\end{aligned}
\tag{2.46}
$$

2.5 Dynamic Equations of Stability for Thin Orthotropic Plates

Differential equations of motion of the plate have been derived on the basis of the Hamilton's principle. It states that the dynamics of a physical system is determined by a variation problem for the functional based on a single function, the Lagrangian, which contains all physical information concerning the system and the forces acting on it. In the dynamic buckling problem, the motion should be understood as the time dependent deflection.

The Hamilton's principle for conservative systems states that the true evolution (compatible with constrains) of the system between two specific states in a specific time range (t_0, t_1) is a stationary point (a point where the variation is zero) of the action functional Ψ. The action functional Ψ is described by the following equation:

$$
\Psi=\int_{t_0}^{t_1}\Lambda dt=\int_{t_0}^{t_1}(K-\Pi)dt, \tag{2.47}
$$

where Λ is the Lagrangian function for the system, K is a kinetic energy of the system, and Π is a total potential energy of the system.

Taking the action functional Ψ in form (2.47), the Hamilton's principle can be written as:

$$\delta\Psi = \delta\int_{t_0}^{t_1}\Lambda dt = \delta\int_{t_0}^{t_1}(K - \Pi)dt = 0. \tag{2.48}$$

The total potential energy variation $\delta\Pi$ for the i-th thin plate (or strip) can be written in the form:

$$\delta\Pi = \delta Q - \delta W, \tag{2.49}$$

where δQ is a variation of the internal elastic strain energy:

$$\delta Q = \int_{\Omega}(\sigma_x\delta\varepsilon_x + \sigma_y\delta\varepsilon_y + \tau_{xy}\delta\gamma_{xy})d\Omega, \tag{2.50}$$

where Ω is a volume of the plate and S is its area, thus the volume can be expressed as $\Omega = l{\cdot}b{\cdot}h$ or $\Omega = S{\cdot}h$.

The variation of the internal elastic strain energy for the i-th plate or strip could be expressed by strain and sectional forces and moments in a following way:

$$\begin{aligned}\delta Q =& \delta Q^m + \delta Q^b\\ =& \int_S(N_x\delta\varepsilon_x + N_y\delta\varepsilon_y + N_{xy}\delta\gamma_{xy})dS - \int_S(M_x\delta w_{,xx} + M_y\delta w_{,yy} + 2M_{xy}\delta w_{,xy})dS.\end{aligned} \tag{2.51}$$

The work W of external forces done on the i-th plate can be expressed as:

$$W = \int_0^b h[p^0(y)u + \tau^0_{xy}(y)v]dy + \int_o^\ell h[p^0(x)v + \tau^0_{xy}(x)u]dx + \int qwdS, \tag{2.52}$$

If the load perpendicular to the plane of the plate (strip) is neglected, (2.52) can be written as:

$$W = \int_0^b h[p^0(y)u + \tau^0_{xy}(y)v]dy + \int_o^\ell h[p^0(x)v + \tau^0_{xy}(x)u]dx, \tag{2.53}$$

where: $p^0(x), p^0(y), \tau^0_{xy}(x), \tau^0_{xy}(y)$ are pre-buckling loads applied to the middle surface of the plate (wall or strip) under consideration.

For thin plates, it is assumed that the displacements u and v do not depend on the rotations $w_{,x}$ and $w_{,y}$ and, therefore, do not depend on the coordinate z. This approach results in the exclusion of the rotational inertia [7] from the equation for kinetic energy which for the i-th thin plate (strip) can be written as:

$$K = \frac{1}{2}\rho \int_{\Omega} \left((\dot{u})^2 + (\dot{v})^2 + (\dot{w})^2 \right) d\Omega, \tag{2.54}$$

where the dot denotes differentiation with respect to time.

The Hamilton's principle, it is a variation of the action functional $\delta\Psi$ (2.48) for the i-th thin plate (strip or wall) after taking into consideration from (2.49) to (2.54) and assuming a constant density for all layers ($\rho = \rho_{\mathrm{k}} = \mathrm{const}$), can be written as:

$$\delta\Psi = \int_{t_0}^{t_1} \left(-\delta K + \delta Q^m + \delta Q^b - \delta W \right) dt = 0 \tag{2.55}$$

where:

$$\begin{aligned}
\int_{t_0}^{t_1} \delta Q^m dt = & \int_{t_0}^{t_1}\int_0^b N_x \delta u dy dt|_{x=const} \\
& - \int_{t_0}^{t_1}\int_S N_{x,x} \delta u dS dt + \int_{t_0}^{t_1}\int_0^b N_x u_{,x} \delta u dy dt|_{x=const} \\
& - \int_{t_0}^{t_1}\int_S (N_x u_{,x})_{,x} \delta u dS dt \int_{t_0}^{t_1}\int_0^b N_x v_{,x} \delta v dy dt|_{x=const} - \int_{t_0}^{t_1}\int_S (N_x v_{,x})_{,x} \delta v dS dt \\
& + \int_{t_0}^{t_1}\int_0^b N_x w_{,x} \delta w dy dt|_{x=const} - \int_{t_0}^{t_1}\int_S (N_x w_{,x})_{,x} \delta w dS dt + \int_{t_0}^{t_1}\int_0^b N_y \delta v dy dt|_{x=const} \\
& - \int_{t_0}^{t_1}\int_S N_{y,y} \delta v dS dt + \int_{t_0}^{t_1}\int_0^l N_y u_{,y} \delta u dx dt|_{y=const} - \int_{t_0}^{t_1}\int_S (N_y u_{,y})_{,y} \delta u dS dt \\
& + \int_{t_0}^{t_1}\int_0^l N_y v_{,y} \delta v dx dt|_{y=const} - \int_{t_0}^{t_1}\int_S (N_y v_{,y})_{,y} \delta v dS dt + \int_{t_0}^{t_1}\int_0^l N_y w_{,y} \delta w dx dt|_{y=const} \\
& - \int_{t_0}^{t_1}\int_S (N_y w_{,y})_{,y} \delta w dS dt + \int_{t_0}^{t_1}\int_0^l N_{xy} \delta u dx dt|_{y=const} - \int_{t_0}^{t_1}\int_S N_{xy,y} \delta u dS dt \\
& + \int_{t_0}^{t_1}\int_0^b N_{xy} \delta v dy dt|_{x=const} - \int_{t_0}^{t_1}\int_S N_{xy,x} \delta v dS dt + \int_{t_0}^{t_1}\int_0^l N_{xy} u_{,x} \delta u dx dt|_{y=const} \\
& - \int_{t_0}^{t_1}\int_S (N_{xy} u_{,x})_{,y} \delta u dS dt + \int_{t_0}^{t_1}\int_0^b N_{xy} u_{,y} \delta u dy dt|_{x=const} - \int_{t_0}^{t_1}\int_S (N_{xy} u_{,y})_{,x} \delta u dS dt
\end{aligned}$$

$$+\int_{t_0}^{t_1}\int_0^l N_{xy}v_{,x}\delta v dxdt|_{y=const}-\int_{t_0}^{t_1}\int_S (N_{xy}v_{,x})_{,y}\delta v dSdt$$
$$+\int_{t_0}^{t_1}\int_0^b N_{xy}v_{,y}\delta v dydt|_{x=const}-\int_{t_0}^{t_1}\int_S (N_{xy}v_{,y})_{,x}\delta v dSdt$$
$$+\int_{t_0}^{t_1}\int_0^l N_{xy}w_{,x}\delta w dxdt|_{y=const}-\int_{t_0}^{t_1}\int_S (N_{xy}w_{,x})_{,y}\delta w dSdt$$
$$+\int_{t_0}^{t_1}\int_0^b N_{xy}w_{,y}\delta w dydt|_{x=const}-\int_{t_0}^{t_1}\int_S (N_{xy}w_{,y})_{,x}\delta w dSdt. \quad (2.56)$$

$$\int_{t_0}^{t_1}\delta Q^b dt = -\int_{t_0}^{t_1}\int_0^b M_x\delta w_x dydt|_{x=const}-\int_{t_0}^{t_1}\int_0^b M_{x,x}\delta w dydt|_{x=const}+2\int_{t_0}^{t_1}\int_S M_{x,xx}\delta w dSdt$$
$$-\int_{t_0}^{t_1}\int_0^l M_y\delta w_y dxdt|_{y=const}-\int_{t_0}^{t_1}\int_0^l M_{y,y}\delta w dxdt|_{y=const}+2\int_{t_0}^{t_1}\int_S M_{y,yy}\delta w dSdt$$
$$-2\int_{t_0}^{t_1}\left[M_{xy}\delta w\right]_{,xy}dt|_{\substack{x=const\\ y=const}}-2\int_{t_0}^{t_1}\int_0^l M_{xy,x}\delta w dxdt|_{y=const}-2\int_{t_0}^{t_1}\int_0^b M_{xy,y}\delta w dydt|_{x=const}$$
$$+2\int_{t_0}^{t_1}\int_S M_{xy,xy}\delta w dSdt. \quad (2.57)$$

$$\int_{t_0}^{t_1}\delta K dt = \int_{t_0}^{t_1}\int_S \rho h\,\ddot{u}\,\delta u dSdt+\int_{t_0}^{t_1}\int_S \rho h\,\ddot{v}\,\delta v dSdt+\int_{t_0}^{t_1}\int_S \rho h\,\ddot{w}\,\delta w dSdt$$
$$-\int_S \rho h\,\dot{u}\,\delta u dS|_{t=const}-\int_S \rho h\,\dot{v}\,\delta v dS|_{t=const}-\int_S \rho h\,\dot{w}\,\delta w dS|_{t=const}. \quad (2.58)$$

$$\int_{t_0}^{t_1}\delta W dt = \int_{t_0}^{t_1}\int_0^b hp^0(y)\delta u dydt|_{x=const}+\int_{t_0}^{t_1}\int_0^b h\tau_{xy}^0(y)\delta v dydt|_{x=const}$$
$$+\int_{t_0}^{t_1}\int_o^\ell hp^0(x)\delta v dxdt|_{y=const}+\int_{t_0}^{t_1}\int_o^\ell h\tau_{xy}^0(x)\delta u dxdt|_{y=const}. \quad (2.59)$$

The Lagrangian function for the whole system is equal to the sum of the Lagrangian functions of all n plates the system is composed of. To determine the variation of action $\delta\Psi$ for the i-th plate, the following identity:

$$X\,\delta Y = \delta(XY) - Y\delta X \tag{2.60}$$

is used.

In the obtained equation, the terms with the same variation have been grouped and then each of the obtained group of terms (due to the mutual independence of variations) has been equated to zero, giving:

- equilibrium equations:

$$\begin{aligned}
&\int_{t_0}^{t_1}\int_S \{[N_{x,x} + N_{xy,y} + (N_x u_{,x})_{,x} + (N_y u_{,y})_{,y} + (N_{xy} u_{,x})_{,y} + (N_{xy} u_{,y})_{,x}] - h\rho\,\ddot{u}\}\delta u dS dt = 0,\\
&\int_{t_0}^{t_1}\int_S \{[N_{xy,x} + N_{y,y} + (N_x v_{,x})_{,x} + (N_y v_{,y})_{,y} + (N_{xy} v_{,x})_{,y} + (N_{xy} v_{,y})_{,x}] - h\rho\,\ddot{v}\}\delta v dS dt = 0,\\
&\int_{t_0}^{t_1}\int_S \{[M_{x,xx} + M_{y,yy} + 2M_{xy,xy} + (N_x w_{,x})_{,x} + (N_y w_{,y})_{,y} + (N_{xy} w_{,x})_{,y} + (N_{xy} w_{,y})_{,x}]\\
&\quad - h\rho\,\ddot{w}\}\delta w dS dt = 0,
\end{aligned} \tag{2.61}$$

- boundary conditions for lateral edges of the plate ($x = \text{const}$):

$$\begin{aligned}
&\int_{t_0}^{t_1}\int_0^b [N_x + N_x u_{,x} + N_{xy} u_{,y} - hp^0(y)]\delta u dy dt|_{x=const} = 0,\\
&\int_{t_0}^{t_1}\int_0^b [N_{xy} + N_x v_{,x} + N_{xy} v_{,y} - h\tau^0_{xy}(y)]\delta v dy dt|_{x=const} = 0,\\
&\int_{t_0}^{t_1}\int_0^b M_x \delta w_{,x} dy dt|_{x=const} = 0,\\
&\int_{t_0}^{t_1}\int_0^b (M_{x,x} + 2M_{xy,y} + N_x w_{,x} + N_{xy} w_{,y})\delta w dy dt|_{x=const} = 0,
\end{aligned} \tag{2.62}$$

- boundary conditions for longitudinal edges of the plate ($y = \text{const}$):

$$\begin{aligned}
&\int_{t_0}^{t_1}\int_0^{\ell}[N_y + N_y v_{,y} + N_{xy} v_{,x} - hp^0(x)]\delta v dxdt\big|_{y=const} = 0,\\
&\int_{t_0}^{t_1}\int_0^{\ell}[N_{xy} + N_y u_{,y} + N_{xy} u_{,x} - h\tau_{xy}^0(x)]\delta u dxdt\big|_{y=const} = 0,\\
&\int_{t_0}^{t_1}\int_0^{\ell} M_y \delta w_{,y} dxdt\big|_{y=const} = 0,\\
&\int_{t_0}^{t_1}\int_0^{\ell}(M_{y,y} + 2M_{xy,x} + N_y w_{,y} + N_{xy} w_{,x})\delta w dxdt\big|_{y=const} = 0,
\end{aligned} \tag{2.63}$$

- boundary conditions for the plate corners (x = const and y = const):

$$\int_{t_0}^{t_1} 2M_{xy}\delta w dt|_{x=const}|_{y=const} = 0, \tag{2.64}$$

- initial conditions for t = const:

$$\begin{aligned}
&\int_S h\rho\dot{u}\delta u dS|_{t=const} = 0,\\
&\int_S h\rho\dot{v}\delta v dS|_{t=const} = 0,\\
&\int_S h\rho\dot{w}\delta w dS|_{t=const} = 0,
\end{aligned} \tag{2.65}$$

which are fulfilled for the entire structure, so if one applies the restrictions at an instant of the initial t_0 and at an instant of the final t_1, then the displacement variations are zero at all points of the structure. Then, system of equations (2.65) vanishes,

- the already used relationship between deformations and internal forces and moments (2.31) or further, up to (2.34) for the orthotropic model of material can be written as follows:

$$\int_{t_0}^{t_1}\int_S \left(E_x h\varepsilon_x - N_x + \nu_{xy}N_y\right)\delta N_x dSdt = 0,$$
$$\int_{t_0}^{t_1}\int_S \left(E_y h\varepsilon_y + \nu_{yx}N_x - N_y\right)\delta N_y dSdt = 0, \tag{2.66}$$
$$\int_{t_0}^{t_1}\int_S \left(2Gh\varepsilon_{xy} - N_{xy}\right)\delta N_{xy} dSdt = 0,$$

$$\int_{t_0}^{t_1}\int_S \left(\frac{E_x h^3}{12}\kappa_x - M_x + \nu_{xy}M_y\right)\delta M_x dSdt = 0,$$
$$\int_{t_0}^{t_1}\int_S \left(\frac{E_y h^3}{12}\kappa_y + \nu_{yx}M_x - M_y\right)\delta M_y dSdt = 0, \tag{2.67}$$
$$\int_{t_0}^{t_1}\int_S \left(\frac{Gh^3}{6}\kappa_{xy} - M_{xy}\right)\delta M_{xy} dSdt = 0.$$

References

1. Kolakowski Z, Kowal-Michalska K (eds) (1999) Selected problems of instabilities in composite structures. A series of monographes, Technical University of Lodz Press, Lodz
2. Krolak M (ed) (1995) Statecznosc, stany zakrytyczne i nosnosc graniczna cienkosciennych dzwigarow o scianach plaskich. Series of monographs. Technical University of Lodz Press, Lodz
3. Kubiak T (2001) Postbuckling behavior of thin-walled girders with orthotropy varying widthwise. Int J Solid Struct 38(28–29):4839–4856
4. Kubiak T (2005) Dynamic buckling of thin-walled composite plates with varying widthwise material properties. Int J Solid Struct 45:5555–5567
5. Kubiak T (2007) Interakcyjne wyboczenie dynamiczne cienkosciennych slupow. Technical University Press, Lodz
6. Pietraszkiewicz W (1989) Geometrically nonlinear theories of thin elastic shells. Adv Mech 12(1):51–130
7. Cz Wozniak (ed) (2001) Mechanika sprezystych plyt i powlok. PWN, Warsaw

Chapter 3
Analytical–Numerical Method

The method proposed and explained below allows one to determine critical loads, natural frequencies and coefficients of the equation describing the postbuckling equilibrium path for thin orthotropic plates or girders, columns and beams composed of flat orthotropic plates (walls). This method also allows one to analyse a dynamic response of the plate structure subjected to pulse loading. Taking the deflections as a function of time and applying the relevant dynamic buckling criteria, it is possible to determine the dynamic critical load.

The analytical–numerical method [9–11, 13, 15, 16, 20–22, 24, 26, 27] developed for many years in the Department of Strength of Materials has been used to solve static and dynamic buckling problems and to analyse the postbuckling behaviour of thin-walled composite structures. The proposed method is based on the asymptotic Koiter [4, 5], Byskov and Hutchinson [2] theory for conservative systems.

The way to find the postbuckling behaviour of plates, girders and beam-columns made of composite materials (multilayered composites) using the asymptotic Koiter theory was explained in [12, 17]. The semi-analytical method proposed by Kolakowski and enabling an approximate analysis of the postbuckling behaviour in terms of the second order nonlinear approximation regarding only a linear analysis consists in the determination of the coefficients describing the postbuckling equilibrium path.

This aim of this monograph is limited to present a method of solution for structures made of iso- or orthotropic materials only.

The most important advantage of this method is that it enables one to describe a complete range of behaviour of thin-walled structures from all global (i.e., flexural, flexural-torsional, lateral, distortional buckling and their combinations) to the local dynamic stability. In the solution obtained, the shear lag phenomenon, the effect of cross-sectional distortions and also the interaction between all walls of the structure are included.

In the previous chapter, the equations describing relations for isotropic, orthotropic and layered composite plates have been derived. The method of

T. Kubiak, *Static and Dynamic Buckling of Thin-Walled Plate Structures*,
DOI: 10.1007/978-3-319-00654-3_3,

solution presented here allows for finding buckling load with the corresponding mode for all the above-mentioned types of materials.

The differential equations of equilibrium for the orthotropic plate or strip can be derived directly from (2.61) and have the form:

$$
\begin{aligned}
&N_{x,x} + N_{xy,y} + \{(N_x u_{,x})_{,x} + (N_y u_{,y})_{,y} + (N_{xy} u_{,x})_{,y} + (N_{xy} u_{,y})_{,x}\} - h\rho\ddot{u} = 0,\\
&N_{xy,x} + N_{y,y} + \{(N_x v_{,x})_{,x} + (N_y v_{,y})_{,y} + (N_{xy} v_{,x})_{,y} + (N_{xy} v_{,y})_{,x}\} - h\rho\ddot{v} = 0,\\
&M_{x,xx} + M_{y,yy} + 2M_{xy,xy} + (N_x w_{,x})_{,x} + (N_y w_{,y})_{,y} + (N_{xy} w_{,x})_{,y} + (N_{xy} w_{,y})_{,x} - h\rho\ddot{w} = 0.
\end{aligned}
\tag{3.1}
$$

The presented above equilibrium equations, after omitting the inertia forces $h\rho\,\ddot{u}$, $h\rho\,\ddot{v}$ and $h\rho\,\ddot{w}$, become the static equilibrium equations for thin plates allowing for analysis of both local and global buckling modes.

As the wave propagation effects have been neglected, the boundary conditions referring to the column simply supported at its both ends, i.e., $x = 0$ and $x = l$, according to (2.62), are assumed to be:

$$
\begin{aligned}
&\frac{1}{b_i}\int N_{ix}(x_i = 0, y_i, t)dy_i = \frac{1}{b_i}\int N_{ix}(x_i = l, y_i, t)dy_i = N_{ix}^{(0)},\\
&v_i(x_i = 0, y_i, t) = v_i(x_i = l, y_i, t) = 0,\\
&w_i(x_i = 0, y_i, t) = w_i(x_i = l, y_i, t) = 0,\\
&M_{ix}(x_i = 0, y_i, t) = M_{iy}(x_i = l, y_i, t) = 0.
\end{aligned}
\tag{3.2}
$$

The condition written as the first one in (3.2) is satisfied for the prebuckling state and the first order approximation, condition (3.2) for the deflection v is satisfied for the first and second order approximations, whereas the other two conditions are met for the prebuckling state as well as for the first and second order approximation. The condition of displacement in the y direction in the prebuckling state can be found for example in [26]. This approach allows for taking into account the impact of the Poisson effect on the edges of column walls. The boundary conditions described by (3.2) assume a lack of displacement in the transverse v and normal w directions to the surface of points lying at the loaded edges of a single plate or a wall of the column. Furthermore, it is assumed that the moments M_{ix} (as a vector parallel to the edge of the plate or the end edge of the column walls) are zero.

For structures with material properties varying widthwise, the strip model which imposes the boundary conditions modification in the second order approximation (see Sect. 6.2) [26] has been adopted. The modification consists in changing the first condition of (3.2) into the following form:

$$
\sum_{i=1}^{J}\frac{1}{b_i}\int_0^{b_i} N_{ix}^{(2)}dy_i\Big|_{x=0;l} = 0.
\tag{3.3}
$$

The summation is performed only for the J number of the strips between which the angle $\varphi_{i,i+1}$ (Fig. 3.1) is equal to zero.

To determine the boundary conditions on longitudinal edges of plates or free edges of columns with open cross-sections, Eq. (2.63) have been used. Directly from (2.65), the following initial conditions result:

$$\begin{aligned}
&\dot{u}_i(x_i, y_i, t = t_0) = \widetilde{u}_i(x_i, y_i) \text{ and } u_i(x_i, y_i, t = t_0) = \overline{u_i}(x_i, y_i),\\
&\dot{v}_i(x_i, y_i, t = t_0) = \widetilde{v}_i(x_i, y_i) \text{ and } v_i(x_i, y_i, t = t_0) = \overline{v_i}(x_i, y_i),\\
&\dot{w}_i(x_i, y_i, t = t_0) = \widetilde{w}_i(x_i, y_i) \text{ and } w_i(x_i, y_i, t = t_0) = \overline{w_i}(x_i, y_i),
\end{aligned} \tag{3.4}$$

where the following functions $\overline{u_i}$, $\overline{v_i}$, $\overline{w_i}$, $\widetilde{u}_i$, $\widetilde{v}_i$, $\widetilde{w_i}$ are given for the initial moment $t = t_0$.

Static and kinematic junction conditions on the longitudinal edges of adjacent plates (Fig. 3.1), according to (2.63), can be written as:

$$\begin{aligned}
&u_{i+1}\big|_{y_{i+1}=0} = u_i\big|_{y_i=b_i},\\
&w_{i+1}\big|_{y_{i+1}=0} = w_i\big|_{y_i=b_i}\cdot\cos(\varphi_{i;i+1}) - v_i\big|_{y_i=b_i}\cdot\sin(\varphi_{i;i+1}),\\
&v_{i+1}\big|_{y_{i+1}=0} = w_i\big|_{y_i=b_i}\sin(\varphi_{i;i+1}) + v_i\big|_{y_i=b_i}\cdot\cos(\varphi_{i;i+1}),\\
&w_{i+1,y}\big|_{y_{i+1}=0} = w_{i,y}\big|_{y_i=b_i},\\
&M_{(i+1)y}\big|_{y_{i+1}=0} = M_{iy}\big|_{y_i=b_i},\\
&N^*_{(i+1)y}\Big|_{y_{i+1}=0} - N^*_{iy}\Big|_{y_i=b_i}\cdot\cos(\varphi_{i;i+1}) - Q^*_{iy}\Big|_{y_i=b_i}\cdot\sin(\varphi_{i;i+1}) = 0,\\
&Q^*_{(i+1)y}\Big|_{y_{i+1}=0} + N^*_{iy}\Big|_{y_i=b_i}\cdot\sin(\varphi_{i;i+1}) - Q^*_{iy}\Big|_{y_i=b_i}\cdot\cos(\varphi_{i;i+1}) = 0,\\
&N^*_{(i+1)xy}\Big|_{y_{i+1}=0} = N^*_{ixy}\Big|_{y_i=b_i},
\end{aligned} \tag{3.5}$$

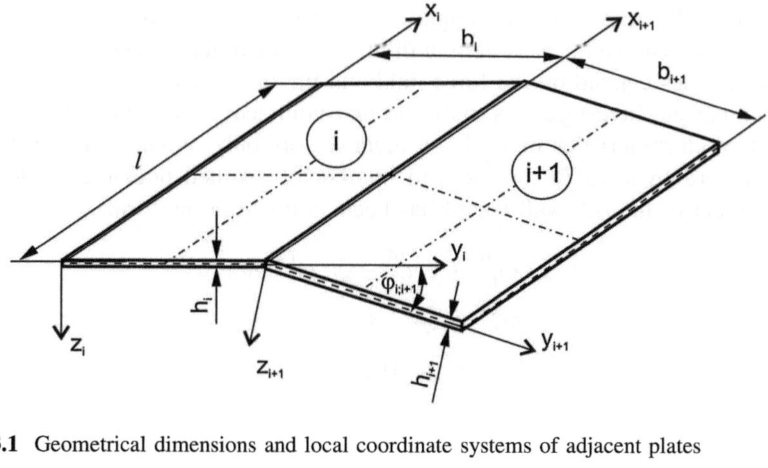

Fig. 3.1 Geometrical dimensions and local coordinate systems of adjacent plates

where:

$$
\begin{aligned}
N_{iy}^{*} &= N_{iy} + N_{iy}v_{i,y} + N_{ixy}v_{i,x}, \\
N_{ixy}^{*} &= N_{ixy} + N_{ixy}u_{i,x} + N_{iy}u_{i,y}, \\
M_{iy} &= -\eta_i D_i (w_{i,yy} + v_i w_{i,xx}), \\
Q_{iy}^{*} &= -\eta_i D_i w_{i,yyy} - (v_i \eta_i D_i + 2D_{1i}) w_{i,xxy} + N_{iy} w_{i,y} + N_{ixy} w_{i,x}.
\end{aligned} \tag{3.6}
$$

According to the Koiter [4], Byskov and Hutchinson [2] theory, the fields of displacements $\overline{U}_i$ and the sectional forces $\overline{N}_i$ have been expanded into power series with respect to the parameter ξ_j, i.e., the linear eigenvector amplitude of buckling (normalised with the equality condition between the maximum deflection and the thickness of the first plate h_1):

$$
\begin{aligned}
\overline{U}_i &= \lambda \overline{U}_i^{(0)} + \xi_j \overline{U}_i^{(j)} + \xi_j \xi_k \overline{U}_i^{(jk)} + \ldots, \\
\overline{N}_i &= \lambda \overline{N}_i^{(0)} + \xi_j \overline{N}_i^{(j)} + \xi_j \xi_k \overline{N}_i^{(jk)} + \ldots,
\end{aligned} \tag{3.7}
$$

where λ is a load parameter, $\overline{U}_i^{(0)}, \overline{N}_i^{(0)}$ are displacement and sectional force fields of the zero approximation (the pre-buckling state), $\overline{U}_i^{(j)}, \overline{N}_i^{(j)}$ are displacement and sectional force fields of the first order approximation (the buckling state), $\overline{U}_i^{(jk)}, \overline{N}_i^{(jk)}$ are second order approximation fields (the post-buckling state). The summation rule after repeating the indices j and k ($j = 1, \ldots, s$ and $k = 1, \ldots, s$, where s is a number of coupled buckling modes) has been assumed.

By substituting expansions (3.7) into equations of equilibrium (3.1) with neglected inertia terms (the static buckling problem), junction conditions (3.5) and boundary conditions (3.2) with modification (3.3), the boundary problem of the zero, first and second order has been obtained (see, for example: [4, 10, 21, 25, 27, 41]). The zero approximation describes the prebuckling state, whereas the first order approximation allows for determination of critical loads and the buckling modes corresponding to them, taking into account minimisation with respect to the number of halfwaves m in the lengthwise direction. The second order approximation is reduced to a linear system of differential heterogeneous equations, whose right-hand sides depend on the force field and the first order displacements only.

In the current investigations, only the first nonlinear approximation, in which the system characteristics depends on eigenvectors only, is taken into consideration. According to (3.7) and the s-th number of coupled buckling modes, the displacement of the i-th wall (strip) has been assumed in the form:

$$
\begin{aligned}
u_i &= \lambda u_i^{(0)} + \xi_j u_i^{(j)} + \xi_j \xi_k u_i^{(jk)} + \ldots, \\
v_i &= \lambda v_i^{(0)} + \xi_j v_i^{(j)} + \xi_j \xi_k v_i^{(jk)} + \ldots, \\
w_i &= \xi_j w_i^{(j)} + \xi_j \xi_k w_i^{(jk)} + \ldots
\end{aligned} \tag{3.8}
$$

According to the explanation given by Kolakowski in [20] and the solution presented in [10], the mixed part $\xi_j\xi_k$ $(j \neq k)$ can be neglected. Thus, after taking into account only the zero, first and second order approximation, (3.8) have the form:

$$
\begin{aligned}
u_i &= \lambda u_i^{(0)} + \xi_j u_i^{(j)} + \xi_j^2 u_i^{(jj)}, \\
v_i &= \lambda v_i^{(0)} + \xi_j v_i^{(j)} + \xi_j^2 v_i^{(jj)}, \\
w_i &= \xi_j w_i^{(j)} + \xi_j^2 w_i^{(jj)}.
\end{aligned} \tag{3.9}
$$

In the presented analytical–numerical method, it has been decided—according to the nonlinear von Kármán and Marquerre plate theory—to take only the following form of geometrical relations:

$$
\begin{aligned}
\varepsilon_{ix} &= u_{i,x} + \frac{1}{2}w_{i,x}^2 + \frac{1}{2}v_{i,x}^2, \\
\varepsilon_{iy} &= v_{i,y} + \frac{1}{2}w_{i,y}^2 + \frac{1}{2}u_{i,x}^2, \\
2\varepsilon_{ixy} &= \gamma_{ixy} = u_{i,y} + v_{i,x} + w_{i,x}w_{i,y}.
\end{aligned} \tag{3.10}
$$

The above simplification has been proven by other authors [6–10, 13–16, 21–23, 29, 30, 32, 33, 35–39] dealing with buckling and the postbuckling behaviour of thin-walled structures subjected to compression or/and bending. Moreover, after a wide range of numerical analyses in the first order approximation [10, 25], it has been noticed that taking into account geometrical relations (2.2) in simplified form (3.10) has no significant influence on the results of calculations. Differences appear only when the global buckling loads are calculated and these differences are in the range of 2–3 %.

The linearly distributed in-plane load (pure bending or eccentricity compression) applied to plates or girder walls has been discretized in such a way that the plate has been divided into strips for which the shortening (load in the longitudinal direction) has a constant value, i.e., each strip under analysis is uniformly compressed (Fig. 3.2). The above explained method of load distribution is necessary in order to apply the transition matrix method which allows for finding the deflection function in the perpendicular direction.

In the pre-buckling state, the displacement for the zero order approximation is assumed as:

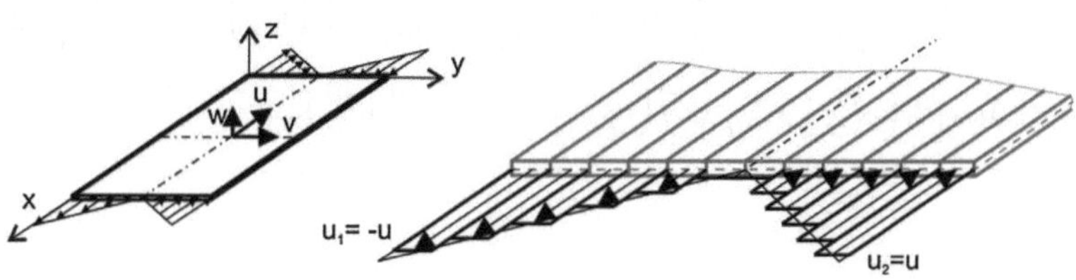

Fig. 3.2 Linear discretization for load as displacement (*pure bending*) for the prebuckling state

$$\begin{aligned} u_i^{(0)} &= \left(\frac{l}{2} - x_i\right)\Delta_i, \\ v_i^{(0)} &= \frac{A_{12i}}{A_{22i}} y_i \Delta_i, \\ w_i^{(0)} &= 0, \end{aligned} \tag{3.11}$$

where Δ_i is an actual load, which is specified as a product of the unit loading system and the scalar load factor. The parameter A_{12i} and A_{22i} are defined by material properties (2.37).

According to the assumed relation from (3.9) to (3.11) and (2.35), the deformation fields for the zero, first and second order can be written as:

$$\begin{aligned} \varepsilon_{ix}^m &= \lambda u_{i,x}^{(0)} + \xi_j u_{i,x}^{(j)} + \xi_j^2 \left(u_{i,x}^{(jj)} + \frac{1}{2}\left(v_{i,x}^{(j)}\right)^2 + \frac{1}{2}\left(w_{i,x}^{(j)}\right)^2\right), \\ \varepsilon_{iy}^m &= \lambda v_{i,y}^{(0)} + \xi_j v_{i,y}^{(j)} + \xi_j^2 \left(v_{i,y}^{(jj)} + \frac{1}{2}\left(u_{i,y}^{(j)}\right)^2 + \frac{1}{2}\left(w_{i,y}^{(j)}\right)^2\right), \\ \gamma_{ixy}^m &= \xi_j\left(u_{i,y}^{(j)} + v_{i,x}^{(j)}\right) + \xi_j^2\left(u_{i,y}^{(jj)} + v_{i,x}^{(jj)} + w_{i,x}^{(j)} w_{i,y}^{(j)}\right), \end{aligned} \tag{3.12}$$

and the change of bending and twisting curvatures of the middle surface are:

$$\begin{aligned} \kappa_{ix} &= -\left(\xi_j w_{i,xx}^{(j)} + \xi_j^2 w_{i,xx}^{(jj)}\right), \\ \kappa_{iy} &= -\left(\xi_j w_{i,yy}^{(j)} + \xi_j^2 w_{i,yy}^{(jj)}\right), \\ \kappa_{ixy} &= -\left(\xi_j w_{i,xy}^{(j)} + \xi_j^2 w_{i,xy}^{(jj)}\right). \end{aligned} \tag{3.13}$$

After assuming relations (3.12), (3.13) and (2.35), the generalized (forces and moments) sectional force fields for the zero, first and second order have the form:

$$\begin{aligned} N_{ix} &= A_{11i}\left\{\lambda\left[u_{i,x}^{(0)} + \frac{A_{12i}}{A_{11i}} v_{i,y}^{(0)}\right] + \xi_j\left[u_{i,x}^{(j)} + \frac{A_{12i}}{A_{11i}} v_{i,y}^{(j)}\right]\right. \\ &\left.+\xi_j^2\left[u_{i,x}^{(jj)} + \frac{1}{2}\left(v_{i,x}^{(j)}\right)^2 + \frac{1}{2}\left(w_{i,x}^{(j)}\right)^2 + \frac{A_{12i}}{A_{11i}}\left(v_{i,y}^{(jj)} + \frac{1}{2}\left(u_{i,y}^{(j)}\right)^2 + \frac{1}{2}\left(w_{i,y}^{(j)}\right)^2\right)\right]\right\}, \\ N_{iy} &= A_{22i}\left\{\xi_j\left[v_{i,y}^{(j)} + \frac{A_{12i}}{A_{22i}} u_{i,x}^{(j)}\right]\right. \\ &\left.+\xi_j^2\left[v_{i,y}^{(jj)} + \frac{1}{2}\left(u_{i,y}^{(j)}\right)^2 + \frac{1}{2}\left(w_{i,y}^{(j)}\right)^2 + \frac{A_{12i}}{A_{22i}}\left(u_{i,x}^{(jj)} + \frac{1}{2}\left(v_{i,x}^{(j)}\right)^2 + \frac{1}{2}\left(w_{i,x}^{(j)}\right)^2\right)\right]\right\}, \\ N_{ixy} &= A_{66i}\left[\xi_j\left(u_{i,y}^{(j)} + v_{i,x}^{(j)}\right) + \xi_j^2\left(u_{i,y}^{(jj)} + v_{i,x}^{(jj)} + w_{i,x}^{(j)} w_{i,y}^{(j)}\right)\right], \end{aligned} \tag{3.14}$$

$$
\begin{aligned}
M_{ix} &= -D_{11i}\left[\xi_j\left(w_{i,xx}^{(j)} + \frac{D_{12i}}{D_{11i}}w_{i,yy}^{(j)}\right) + \xi_j^2\left(w_{i,xx}^{(jj)} + \frac{D_{12i}}{D_{11i}}w_{i,yy}^{(jj)}\right)\right],\\
M_{iy} &= -D_{22i}\left[\xi_j\left(w_{i,yy}^{(j)} + \frac{D_{12i}}{D_{22i}}w_{i,xx}^{(j)}\right) + \xi_j^2\left(w_{i,yy}^{(jj)} + \frac{D_{12i}}{D_{22i}}w_{i,xx}^{(jj)}\right)\right],\\
M_{ixy} &= -D_{66i}\left(\xi_j w_{i,xy}^{(j)} + \xi_j^2 w_{i,xy}^{(jj)}\right).
\end{aligned}
\tag{3.15}
$$

3.1 Static Buckling and Postbuckling Behaviour

The equilibrium equations for the i-th thin composite plate or strip subjected to static load for solving the buckling and postbuckling behaviour according to (3.1) have the form:

$$
\begin{aligned}
&N_{x,x} + N_{xy,y} + \left(N_y u_{,y}\right)_{,y} = 0,\\
&N_{xy,x} + N_{y,y} + \left(N_x v_{,x}\right)_{,x} = 0,\\
&M_{x,xx} + M_{y,yy} + 2M_{xy,xy} + \left(N_x w_{,x}\right)_{,x} + \left(N_y w_{,y}\right)_{,y} + \left(N_{xy} w_{,x}\right)_{,y} + \left(N_{xy} w_{,y}\right)_{,x} = 0.
\end{aligned}
\tag{3.16}
$$

In all equations in this section, the index i denoting the i-th plate or strip will be omitted.

Taking into account relations (2.44) and (3.12) to (3.15), equilibrium equations (3.16) can be transformed into the equilibrium equations expressed in terms of displacements. The displacement equilibrium equations for the first order can be written as follows:

$$
\begin{aligned}
&u_{,xx}^{(j)} + \frac{A_{12}}{A_{11}}v_{,xy}^{(j)} + \frac{A_{66}}{A_{11}}\left(u_{,yy}^{(j)} + v_{,xy}^{(j)}\right) = 0,\\
&v_{,yy}^{(j)} + \frac{A_{12}}{A_{22}}u_{,xy}^{(j)} + \frac{A_{66}}{A_{22}}\left(u_{,xy}^{(j)} + v_{,xx}^{(j)}\right) + \frac{A_{11}}{A_{22}}\lambda\left(u_{,x}^{(0)} + \frac{A_{12}}{A_{11}}v_{,y}^{(0)}\right)v_{,xx}^{(j)} = 0,\\
&\lambda\left(u_{,x}^{(0)} + \frac{A_{12}}{A_{11}}v_{,y}^{(0)}\right)w_{,xx}^{(j)}\\
&-\left[\frac{D_{11}}{A_{11}}\left(w_{,xxxx}^{(j)} + \frac{D_{12}}{D_{11}}w_{,xxyy}^{(j)}\right) + \frac{D_{22}}{A_{11}}\left(w_{,yyyy}^{(j)} + \frac{D_{12}}{D_{22}}w_{,xxyy}^{(j)}\right) + 4\frac{D_{66}}{A_{11}}w_{,xxyy}^{(j)}\right] = 0.
\end{aligned}
\tag{3.17}
$$

In order to simplify the solution, the dimensionless coordinates dependent on the width of the i-th plate are introduced in the following form:

$$
\zeta = \frac{x}{b},\ \chi = \frac{y}{b}. \tag{3.18}
$$

Numerical aspects of solution convergences for the problem being solved for the first order fields have resulted in an introduction of orthogonal functions in the sense of boundary conditions for two longitudinal edges (2.63). These new orthogonal functions are as follows:

$$\begin{aligned}
&\bar{a} = v^{(j)}_{,\chi} + \frac{A_{12}}{A_{22}} u^{(j)}_{,\zeta}, && \bar{b} = u^{(j)}_{,\chi} + v^{(j)}_{,\zeta}, \\
&\bar{c} = u^{(j)}, && \bar{d} = v^{(j)}, \\
&\bar{e} = w^{(j)}, && \bar{f} = w^{(j)}_{,\chi} \\
&\bar{g} = w^{(j)}_{,\chi\chi} + \frac{D_{12}}{D_{22}} w^{(j)}_{,\zeta\zeta}, && \bar{h} = \frac{D_{22}}{D_{66}}\left(w^{(j)}_{,\chi\chi\chi} + \frac{D_{12}}{D_{22}} w^{(j)}_{,\zeta\zeta\chi}\right) + 4w^{(j)}_{,\zeta\zeta\chi},
\end{aligned} \tag{3.19}$$

where relations (3.18) have been implemented.

The functions $\bar{a}$, $\bar{b}$, $\bar{g}$, $\bar{h}$ correspond to generalized sectional forces (3.14) and (3.15) for the first order approximation.

Taking into account introduced orthogonal functions (3.19) in (3.17), the following homogeneous system of differential equations is obtained:

$$\begin{cases}
\bar{a}_{,\chi} = -\frac{A_{66}}{A_{22}} \bar{b}_{,\zeta} + \lambda\Delta \frac{A_{11}}{A_{22}} \left(1 - \frac{A_{12}}{A_{11}}\frac{A_{12}}{A_{22}}\right) \bar{d}_{,\zeta\zeta}, \\
\bar{b}_{,\chi} = -\frac{A_{11}}{A_{66}} \left(\bar{c}_{,\zeta\zeta} + \frac{A_{12}}{A_{11}} \bar{d}_{,\zeta\chi}\right), \\
\bar{c}_{,\chi} = \bar{b} - \bar{d}_{,\zeta}, \\
\bar{d}_{,\chi} = \bar{a} - \frac{A_{12}}{A_{22}} \bar{c}_{,\zeta}, \\
\bar{e}_{,\chi} = \bar{f}, \\
\bar{f}_{,\chi} = \bar{g} - \frac{D_{12}}{D_{22}} \bar{e}_{,\zeta\zeta}, \\
\bar{g}_{,\chi} = \frac{D_{66}}{D_{22}} (\bar{h} - 4\bar{f}_{,\zeta\zeta}), \\
\bar{h}_{,\chi} = \frac{A_{11}}{D_{66}} \left[-\lambda\Delta\left(1 - \frac{A_{12}}{A_{11}}\frac{A_{12}}{A_{22}}\right) \bar{e}_{,\zeta\zeta} - \frac{D_{11}}{A_{11}} \bar{e}_{,\zeta\zeta\zeta\zeta} - \frac{D_{12}}{A_{11}} \bar{f}_{,\zeta\zeta\chi}\right],
\end{cases} \tag{3.20}$$

The solution to differential equations (3.20) is predicted in the form:

$$\begin{aligned}
&\bar{a} = \overline{A}(\chi) \sin\frac{m\pi b}{l}\zeta, && \bar{b} = \overline{B}(\chi) \cos\frac{m\pi b}{l}\zeta, \\
&\bar{c} = \overline{C}(\chi) \cos\frac{m\pi b}{l}\zeta, && \bar{d} = \overline{D}(\chi) \sin\frac{m\pi b}{l}\zeta, \\
&\bar{e} = \overline{E}(\chi) \sin\frac{m\pi b}{l}\zeta, && \bar{f} = \overline{F}(\chi) \sin\frac{m\pi b}{l}\zeta, \\
&\bar{g} = \overline{G}(\chi) \sin\frac{m\pi b}{l}\zeta, && \bar{h} = \overline{H}(\chi) \sin\frac{m\pi b}{l}\zeta,
\end{aligned} \tag{3.21}$$

where $\overline{A}, \overline{B}, \overline{C}, \overline{D}, \overline{E}, \overline{F}, \overline{G}, \overline{H}$ are initially unknown functions (with the m-harmonics) defining deflection in the transverse direction. These functions are numerically designated using the modified numerical transition matrix method.

Substituting the assumed solutions (3.21) into (3.20), the following system of ordinary differential equations has been obtained:

$$\overline{A}_{,\chi} = \frac{A_{66}}{A_{22}}\left(\frac{m\pi b}{l}\right)\overline{B} - \lambda\Delta\frac{A_{11}}{A_{22}}\left(1 - \frac{A_{12}A_{12}}{A_{11}A_{22}}\right)\left(\frac{m}{l}\right)^2\overline{D},$$

$$\overline{B}_{,\chi} = \frac{A_{11}}{A_{66}}\left[\left(\frac{m\pi b}{l}\right)^2\overline{C} - \frac{A_{12}}{A_{11}}\left(\frac{m\pi b}{l}\right)\overline{D}_{,\chi}\right],$$

$$\overline{C}_{,\chi} = \overline{B} - \left(\frac{m\pi b}{l}\right)\overline{D},$$

$$\overline{D}_{,\chi} = \overline{A} + \frac{A_{12}}{A_{22}}\left(\frac{m\pi b}{l}\right)\overline{C},$$

$$\overline{E}_{,\chi} = \overline{F},$$

$$\overline{F}_{,\chi} = \overline{G} + \frac{D_{12}}{D_{22}}\left(\frac{m\pi b}{l}\right)^2\overline{E},$$

$$\overline{G}_{,\chi} = \frac{D_{66}}{D_{22}}\left[\overline{H} + 4\left(\frac{m\pi b}{l}\right)^2\overline{F}\right],$$

$$\overline{H}_{,\chi} = \frac{A_{11}}{D_{66}}\left(\frac{m\pi b}{l}\right)^2\left[\lambda\Delta\left(1 - \frac{A_{12}A_{12}}{A_{11}A_{22}}\right) - \frac{D_{11}}{A_{11}}\left(\frac{m\pi b}{l}\right)^2\right]\overline{E} + \frac{D_{12}}{A_{11}}\left(\frac{m\pi b}{l}\right)^2\overline{F}_{,\chi}. \tag{3.22}$$

Equilibrium equations, expressed in terms of displacements, for the second order approximation have been obtained after substituting (2.44) and (3.12) to (3.15) into (3.16) and they can be written in the following form:

$$u^{(jj)}_{,xx} + \frac{A_{12}}{A_{11}}v^{(jj)}_{,xy} + \frac{A_{66}}{A_{11}}\left(u^{(jj)}_{,yy} + v^{(jj)}_{,xy}\right) = -R_1,$$

$$v^{(jj)}_{,yy} + \frac{A_{12}}{A_{22}}u^{(jj)}_{,xy} + \frac{A_{66}}{A_{22}}\left(u^{(jj)}_{,xy} + v^{(jj)}_{,xx}\right) + \frac{A_{11}}{A_{22}}\lambda\left(u^{(0)}_{,x} + \frac{A_{12}}{A_{11}}v^{(0)}_{,y}\right)v^{(jj)}_{,xx} = -R_2,$$

$$\lambda\left(u^{(0)}_{,x} + \frac{A_{12}}{A_{11}}v^{(0)}_{,y}\right)w^{(jj)}_{,xx} +$$

$$-\left[\frac{D_{11}}{A_{11}}\left(w^{(jj)}_{,xxxx} + \frac{D_{12}}{D_{11}}w^{(jj)}_{,xxyy}\right) + \frac{D_{22}}{A_{11}}\left(w^{(JJ)}_{,yyyy} + \frac{D_{12}}{D_{22}}w^{(JJ)}_{,xxyy}\right) + 4\frac{D_{66}}{A_{11}}w^{(JJ)}_{,xxyy}\right] = -R_3, \tag{3.23}$$

where the right-hand sides of (3.23) (called “loading” terms by some authors [10]), denoted as R_1, R_2 and R_3, are functions of the first order displacement. They are as follows:

$$\begin{aligned}
R_1 &= v^{(j)}_{,x} v^{(j)}_{,xx} + w^{(j)}_{,x} w^{(j)}_{,xx} + \frac{A_{12}}{A_{11}}\left(u^{(j)}_{,y} u^{(j)}_{,xy} + w^{(j)}_{,y} w^{(j)}_{,xy}\right) + \frac{A_{66}}{A_{11}}\left(w^{(j)}_{,x} w^{(j)}_{,y}\right)_{,y} \\
&\quad + \frac{A_{22}}{A_{11}}\left[\left(v^{(j)}_{,y} + \frac{A_{12}}{A_{22}} u^{(j)}_{,x}\right) u^{(j)}_{,y}\right]_{,y}, \\
R_2 &= u^{(j)}_{,y} u^{(j)}_{,yy} + w^{(j)}_{,y} w^{(j)}_{,yy} + \frac{A_{12}}{A_{22}}\left(v^{(j)}_{,x} v^{(j)}_{,xy} + w^{(j)}_{,x} w^{(j)}_{,xy}\right) + \frac{A_{66}}{A_{22}}\left(w^{(j)}_{,x} w^{(j)}_{,y}\right)_{,x} \\
&\quad + \frac{A_{11}}{A_{22}}\left[\left(u^{(j)}_{,x} + \frac{A_{12}}{A_{11}} v^{(j)}_{,y}\right) v^{(j)}_{,x}\right]_{,x}, \\
R_3 &= A_{11}\left(u^{(j)}_{,x} + \frac{A_{12}}{A_{11}} v^{(j)}_{,y}\right) w^{(jj)}_{,xx} + A_{22}\left(v^{(j)}_{,y} + \frac{A_{12}}{A_{22}} u^{(j)}_{,x}\right) w^{(jj)}_{,yy} \\
&\quad + 2A_{66}\left(u^{(j)}_{,y} + v^{(j)}_{,x}\right) w^{(jj)}_{,xy} - A_{11}\lambda\left(u^{(0)}_{,x} + \frac{A_{12}}{A_{11}} v^{(0)}_{,y}\right) v^{(j)}_{,xx} w^{(j)}_{,y}.
\end{aligned} \tag{3.24}$$

New orthogonal functions for the second order fields are also introduced in the same way as for the first order and they are:

$$\begin{aligned}
\hat{a} &= v^{(jj)}_{,\chi} + \frac{A_{12}}{A_{22}} u^{(jj)}_{,\zeta}, & \hat{b} &= u^{(jj)}_{,\chi} + v^{(jj)}_{,\zeta}, \\
\hat{c} &= u^{(jj)}, & \hat{d} &= v^{(jj)}, \\
\hat{e} &= w^{(jj)}, & \hat{f} &= w^{(jj)}_{,\chi}, \\
\hat{g} &= w^{(jj)}_{,\chi\chi} + \frac{D_{12}}{D_{22}} w^{(jj)}_{,\zeta\zeta}, & \hat{h} &= \frac{D_{22}}{D_{66}}\left(w^{(jj)}_{,\chi\chi\chi} + \frac{D_{12}}{D_{22}} w^{(jj)}_{,\zeta\zeta\chi}\right) + 4w^{(jj)}_{,\zeta\zeta\chi}.
\end{aligned} \tag{3.25}$$

Substituting the functions of the first (3.19) and second (3.25) order approximations to equilibrium equations (3.23), the following linear system of differential equations for the second order approximation is obtained:

$$\begin{cases} \hat{a}_{,\chi} = -\dfrac{A_{66}}{A_{22}}\hat{b}_{,\zeta} + \lambda\Delta\dfrac{A_{11}}{A_{22}}\left(1-\dfrac{A_{12}A_{12}}{A_{11}A_{22}}\right)\hat{d}_{,\zeta\zeta} - \bar{c}_{,\chi}\bar{c}_{,\chi\chi} - \bar{f}\bar{f}_{,\chi} - \dfrac{A_{66}}{A_{22}}\left(\bar{e}_{,\zeta\zeta}\bar{f} + \bar{e}_{,\zeta}\bar{f}_{,\zeta}\right) \\ \quad -\dfrac{A_{12}}{A_{22}}\left(2\bar{d}_{,\zeta}\bar{d}_{,\zeta\chi} + \bar{e}_{,\zeta}\bar{f}_{,\zeta} + \bar{d}_{,\chi}\bar{d}_{,\zeta\zeta}\right) - \dfrac{A_{11}}{A_{22}}\left(\bar{c}_{,\zeta\zeta}\bar{d}_{,\zeta} + \bar{c}_{,\zeta}\bar{d}_{,\zeta\zeta}\right), \\ \hat{b}_{,\chi} = -\dfrac{A_{11}}{A_{66}}\left(\hat{c}_{,\zeta\zeta} + \dfrac{A_{12}}{A_{11}}\hat{d}_{,\zeta\chi}\right) - \bar{d}_{,\zeta}\bar{d}_{,\zeta\zeta} - \bar{e}_{,\zeta}\bar{e}_{,\zeta\zeta} - \dfrac{A_{12}}{A_{11}}\bar{c}_{,\zeta\chi}\bar{c}_{,\chi} - \dfrac{A_{12}+A_{66}}{A_{11}}\bar{f}\bar{f}_{,\zeta} \\ \quad -\bar{e}_{,\zeta}\bar{f}_{,\chi} - \dfrac{A_{22}}{A_{11}}\left(\bar{a}\bar{c}_{,\chi}\right)_{,\chi}, \\ \hat{c}_{,\chi} = \hat{b} - \hat{d}_{,\zeta}, \\ \hat{d}_{,\chi} = \hat{a} - \frac{A_{12}}{A_{22}}\hat{c}_{,\zeta}, \\ \hat{e}_{,\chi} = \hat{f}, \\ \hat{f}_{,\chi} = \hat{g} - \frac{D_{12}}{D_{22}}\hat{e}_{,\zeta\zeta}, \\ \hat{g}_{,\chi} = \frac{D_{66}}{D_{22}}\left(\hat{h} - 4\hat{f}_{,\zeta\zeta}\right), \\ \hat{h}_{,\chi} = \dfrac{A_{11}}{D_{66}}\left[-\lambda\Delta\left(1-\dfrac{A_{12}A_{12}}{A_{11}A_{22}}\right)\hat{e}_{,\zeta\zeta} - \dfrac{D_{11}}{A_{11}}\hat{e}_{,\zeta\zeta\zeta\zeta} - \dfrac{D_{12}}{A_{11}}\hat{f}_{,\zeta\zeta\chi}\right] \\ \quad -\left(A_{11} + A_{12}\dfrac{A_{12}}{A_{22}}\right)\bar{c}_{,\zeta}\bar{e}_{,\zeta\zeta} - A_{12}\bar{a}\bar{e}_{,\zeta\zeta} - A_{22}\bar{a}\bar{f}_{,\chi} - 2A_{66}\bar{b}\bar{f}_{,\zeta} \\ \quad -A_{11}\left(1-\dfrac{A_{12}A_{12}}{A_{11}A_{22}}\right)\lambda\Delta\bar{d}_{,\zeta\zeta}\bar{f}. \end{cases} \tag{3.26}$$

The solution to (3.26) for the second order approximations has been adopted in the following form:

$$\begin{aligned} \hat{a} &= \sum_n \hat{A}_n(\chi)\sin\frac{n\pi b}{l}\zeta + A^*, & \hat{b} &= \sum_n \hat{B}_n(\chi)\cos\frac{n\pi b}{l}\zeta, \\ \hat{c} &= \sum_n \hat{C}_n(\chi)\cos\frac{n\pi b}{l}\zeta + C^*\left(\frac{l}{2} - \zeta b\right), & \hat{d} &= \sum_n \hat{D}_n(\chi)\sin\frac{n\pi b}{l}\zeta, \\ \hat{e} &= \sum_n \hat{E}_n(\chi)\sin\frac{n\pi b}{l}\zeta, & \hat{f} &= \sum_n \hat{F}_n(\chi)\sin\frac{n\pi b}{l}\zeta, \\ \hat{g} &= \sum_n \hat{G}_n(\chi)\sin\frac{n\pi b}{l}\zeta, & \hat{h} &= \sum_n \hat{H}_n(\chi)\sin\frac{n\pi b}{l}\zeta, \end{aligned} \tag{3.27}$$

where: $\hat{A}_n, \hat{B}_n, \hat{C}_n, \hat{D}_n, \hat{E}_n, \hat{F}_n, \hat{G}_n, \hat{H}_n$ are unknown functions, which are determined during the numerical calculations in the second order approximation using the transition matrix method; A^*, C^*—constant values designated for boundary conditions for second order approximation (3.2) with modification (3.3) [26].

The simplicity of orthogonality conditions is one of the main reasons to select the way of solution to the second order approximation by employing the form of series. The orthogonality condition of the fields of the first and second order shows a change in the amplitude of only one harmonic for the second order.

After substituting the expected solution (3.27) to (3.26) and taking into account:

$$
\begin{aligned}
\sin 2\frac{m\pi b}{l}\zeta &= \sum \mathbf{a}_n \cos\frac{n\pi b}{l}\zeta,\\
1 &= \sum \mathbf{b}_n \sin\frac{n\pi b}{l}\zeta,\\
\cos 2\frac{m\pi b}{l}\zeta &= \sum \mathbf{d}_n \sin\frac{n\pi b}{l}\zeta,
\end{aligned}
\tag{3.28}
$$

where:

$$
\begin{aligned}
\mathbf{a}_n &= \frac{2}{\pi}\left(\frac{1}{2m+n}+\frac{1}{2m-n}\right),\\
\mathbf{b}_n &= \frac{4}{\pi n}, \qquad\qquad \mathrm{n} = 1,3,5,7,\ldots,\\
\mathbf{d}_n &= \frac{2}{\pi}\left(\frac{1}{n+2m}+\frac{1}{n-2m}\right),
\end{aligned}
\tag{3.29}
$$

the following system of ordinary differential equations is obtained:

$$
\begin{aligned}
\hat{A}_{n,\chi} =& \frac{A_{66}}{A_{22}}\left(\frac{m\pi b}{l}\right)\hat{B}_n - \lambda\Delta\frac{A_{11}}{A_{22}}\left(1-\frac{A_{12}A_{12}}{A_{11}A_{22}}\right)\left(\frac{m}{l}\right)^2\hat{D}_n\\
&-\frac{1}{2b}\{\mathbf{b}_n\}\left[\bar{C}_{,\chi}\bar{C}_{,\chi\chi}+\bar{F}\bar{F}_{,\chi}+\left(2\frac{A_{66}}{A_{22}}+\frac{A_{12}}{A_{22}}\right)\left(\frac{m\pi b}{l}\right)^2\bar{E}\bar{F}+3\frac{A_{12}}{A_{22}}\left(\frac{m\pi b}{l}\right)^2\bar{D}\bar{D}_{,\chi}\right]\\
&-\frac{1}{2b}\{\mathbf{d}_n\}\left[\bar{C}_{,\chi}\bar{C}_{,\chi\chi}-\bar{F}\bar{F}_{,\chi}+\left(2\frac{A_{66}}{A_{22}}+\frac{A_{12}}{A_{22}}\right)\left(\frac{m\pi b}{l}\right)^2\bar{E}\bar{F}+3\frac{A_{12}}{A_{22}}\left(\frac{m\pi b}{l}\right)^2\bar{D}\bar{D}_{,\chi}\right.\\
&\left.-\frac{A_{11}}{A_{22}}\left(\frac{m\pi b}{l}\right)^3\bar{C}\bar{D}\right],
\end{aligned}
$$

$$
\begin{aligned}
\hat{B}_{n,\chi} &= \tfrac{A_{11}}{A_{66}}\left[\left(\tfrac{m\pi b}{l}\right)^2\hat{C}_n - \tfrac{A_{12}}{A_{11}}\left(\tfrac{m\pi b}{l}\right)\hat{D}_{n,\chi}\right]+\\
&\quad+\tfrac{1}{2b}\{\mathbf{a}_n\}\left[\left(\tfrac{m\pi b}{l}\right)^3\bar{D}^2+\left(\tfrac{m}{l}\right)^3\bar{E}^2+\tfrac{A_{12}}{A_{11}}\left(\tfrac{m\pi b}{l}\right)\left(\bar{C}_{,\chi}\right)^2-\left(\tfrac{m\pi b}{l}\right)\bar{E}\bar{F}\right.\\
&\quad\left.-\tfrac{A_{12}+A_{66}}{A_{11}}\left(\tfrac{m\pi b}{l}\right)\bar{F}^2-\tfrac{A_{22}}{A_{11}}\left(\bar{A}\bar{C}_{,\chi}\right)_{,\chi}\right],\\
\hat{C}_{n,\chi} &= \hat{B}_n - \left(\tfrac{m\pi b}{l}\right)\hat{D}_n,
\end{aligned}
\tag{3.30}
$$

$$\hat{D}_{n,\chi} = \hat{A}_n + \frac{A_{12}}{A_{22}} \left(\frac{m\pi b}{l}\right) \hat{C}_n,$$

$$\hat{E}_{n,\chi} = \hat{F}_n,$$

$$\hat{F}_{n,\chi} = \hat{G}_n + \frac{D_{12}}{D_{22}} \left(\frac{m\pi b}{l}\right)^2 \hat{E}_n,$$

$$\hat{G}_{n,\chi} = \frac{D_{66}}{D_{22}} \left[\hat{H}_n + 4\left(\frac{m\pi b}{l}\right)^2 \hat{F}_n \right],$$

$$\begin{aligned}
\hat{H}_{n,\chi} =& \frac{A_{11}}{D_{66}} \left(\frac{m}{l}\right)^2 \left[\lambda\Delta\left(1 - \frac{A_{12}A_{12}}{A_{11}A_{22}}\right) - \frac{D_{11}}{A_{11}} \left(\frac{m\pi b}{l}\right)^2 \right] \hat{E}_n + \frac{D_{12}}{A_{11}} \left(\frac{m}{l}\right)^2 \hat{F}_{n,\chi} \\
&+ \frac{1}{2b}\{\mathbf{b}_n\} \left[\left(A_{11} + A_{12}\frac{A_{12}}{A_{22}}\right) \left(\frac{m\pi b}{l}\right)^3 \bar{C}\bar{E} - A_{12}\left(\frac{m\pi b}{l}\right)^2 \bar{A}\bar{E} + A_{22}\bar{A}\bar{F}_{,\chi} \right. \\
&\left. + 2A_{66}\left(\frac{m\pi b}{l}\right) \bar{B}\bar{F} - A_{11}\left(1 - \frac{A_{12}A_{12}}{A_{11}A_{22}}\right) \lambda\Delta\left(\frac{m\pi b}{l}\right)^2 \bar{D}\bar{F} \right] \\
&- \frac{1}{2b}\{\mathbf{d}_n\} \left[\left(A_{11} + A_{12}\frac{A_{12}}{A_{22}}\right) \left(\frac{m\pi b}{l}\right)^3 \bar{C}\bar{E} - A_{12}\left(\frac{m\pi b}{l}\right)^2 \bar{A}\bar{E} + A_{22}\bar{A}\bar{F}_{,\chi} \right. \\
&\left. - 2A_{66}\left(\frac{m\pi b}{l}\right) \bar{B}\bar{F} - A_{11}\left(1 - \frac{A_{12}A_{12}}{A_{11}A_{22}}\right) \lambda\Delta\left(\frac{m\pi b}{l}\right)^2 \bar{D}\bar{F} \right].
\end{aligned}$$

Having found the solutions to the first and second order of the boundary problem and taking into consideration (3.9), the coefficients a_{ijs}, b_{ssss} have been determined [2, 21, 25, 26]:

$$\begin{aligned}
a_{ijs} &= \frac{\boldsymbol{\sigma}^{(i)} * L_{11}(\mathbf{U}^{(j)}, \mathbf{U}^{(s)}) + 0.5\boldsymbol{\sigma}^{(s)} * L_{11}(\mathbf{U}^{(i)}, \mathbf{U}^{(j)})}{-\lambda_s \boldsymbol{\sigma}^{(0)} * L_2(\mathbf{U}^{(s)})}, \\
b_{ssss} &= \frac{2\boldsymbol{\sigma}^{(s)} * L_{11}(\mathbf{U}^{(ss)}, \mathbf{U}^{(s)}) + \boldsymbol{\sigma}^{(ss)} * L_{11}(\mathbf{U}^{(s)}, \mathbf{U}^{(s)})}{-\lambda_s \boldsymbol{\sigma}^{(0)} * L_2(\mathbf{U}^{(s)})},
\end{aligned} \tag{3.31}$$

where: λ_s—is the critical load corresponding to the s-th mode, L_{11} is a bilinear operator, L_2 is a quadratic operator, and $\sigma^{(i)}$, $\sigma^{(ij)}$ are stress field tensors in the first and second order.

The postbuckling static equilibrium paths for coupled buckling can be described by the equation:

$$\left(1 - \frac{\lambda}{\lambda_s}\right)\xi_s + a_{ijs}\xi_i\xi_j + b_{ssss}\xi_s^3 = \frac{\lambda}{\lambda_s}\xi_s^*; \ (\mathrm{s} = 1, \ldots, N), \tag{3.32}$$

which for the uncoupled problem has the form:

$$\left(1 - \frac{\lambda}{\lambda_{cr}}\right)\xi + a_{111}\xi^2 + b_{1111}\xi^3 = \xi^* \frac{\lambda}{\lambda_{cr}}, \tag{3.33}$$

where λ_{cr} is a critical load value and ξ_s^*, ξ^* are dimensionless initial imperfections corresponding to the s-th or first buckling mode (an amplitude of the initial imperfection divided by the plate thickness).

In a special case, i.e., for the so-called ideal structure without initial imperfections ($\xi^* = 0$) and when the equilibrium path (a_{111}) is symmetrical, the post-buckling equilibrium path is defined by the equation:

$$\frac{\lambda}{\lambda_{cr}} = 1 + b_{1111}\xi^2. \tag{3.34}$$

3.2 Natural Frequencies

The determination of natural frequencies is similar to the determination of critical buckling load and it requires the eigenvalue problem to be solved.

Natural frequencies of thin-walled structures were determined by solving the dynamic problem, which uses the approach proposed by Koiter in his asymptotic stability theory of conservative systems in the first order approximation [4].

To determine natural frequencies [40] of the structure, the equilibrium equations are in the following form:

$$\begin{aligned}
&N_{x,x} + N_{xy,y} + (N_y u_{,y})_{,y} - h\rho\,\ddot{u} = 0,\\
&N_{xy,x} + N_{y,y} + (N_x v_{,x})_{,x} - h\rho\,\ddot{v} = 0,\\
&M_{x,xx} + M_{y,yy} + 2M_{xy,xy} + (N_x w_{,x})_{,x} + (N_y w_{,y})_{,y} + (N_{xy} w_{,x})_{,y} + (N_{xy} w_{,y})_{,x} - h\rho\,\ddot{w} = 0.
\end{aligned} \tag{3.35}$$

They contain cross-sectional inertia forces acting in the direction normal to the middle surface of the plate (column wall) and in the middle plane of the plate (i.e., $h\rho\,\ddot{u} \neq 0$ and $h\rho\,\ddot{v} \neq 0$).

In all equations in this section, the index i denoting the i-th plate or strip is omitted.

Similarly as in the analysis of stability (Sect. 3.1), relations (2.44) and (3.12) to (3.15) have been taken into account in equilibrium equations (3.35). Therefore, the linear equations of motion expressed in terms of displacement have the form:

$$
\begin{aligned}
&u^{(j)}_{,xx} + \frac{A_{12}}{A_{11}} v^{(j)}_{,xy} + \frac{A_{66}}{A_{11}} \left(u^{(j)}_{,yy} + v^{(j)}_{,xy} \right) - h\rho \ddot{u}^{(j)} = 0,\\
&v^{(j)}_{,yy} + \frac{A_{12}}{A_{22}} u^{(j)}_{,xy} + \frac{A_{66}}{A_{22}} \left(u^{(j)}_{,xy} + v^{(j)}_{,xx} \right) + \frac{A_{11}}{A_{22}} \lambda \left(u^{(0)}_{,x} + \frac{A_{12}}{A_{11}} v^{(0)}_{,y} \right) v^{(j)}_{,xx} - h\rho \ddot{v}^{(j)} = 0,\\
&\lambda \left(u^{(0)}_{,x} + \frac{A_{12}}{A_{11}} v^{(0)}_{,y} \right) w^{(j)}_{,xx} +\\
&- \left[\frac{D_{11}}{A_{11}} \left(w^{(j)}_{,xxxx} + \frac{D_{12}}{D_{11}} w^{(j)}_{,xxyy} \right) + \frac{D_{22}}{A_{11}} \left(w^{(j)}_{,yyyy} + \frac{D_{12}}{D_{22}} w^{(j)}_{,xxyy} \right) + 4 \frac{D_{66}}{A_{11}} w^{(j)}_{,xxyy} \right] - h\rho \ddot{w}^{(j)} = 0.
\end{aligned}
\tag{3.36}
$$

The new orthogonal functions in the sense of boundary conditions for two longitudinal edges (2.63) are introduced and they are as follows:

$$
\begin{aligned}
&\bar{a} = v^{(j)}_{,\chi} + \frac{A_{12}}{A_{22}} u^{(j)}_{,\zeta}, && \bar{b} = u^{(j)}_{,\chi} + v^{(j)}_{,\zeta},\\
&\bar{c} = u^{(j)}, && \bar{d} = v^{(j)},\\
&\bar{e} = w^{(j)}, && \bar{f} = w^{(j)}_{,\chi},\\
&\bar{g} = w^{(j)}_{,\chi\chi} + \frac{D_{12}}{D_{22}} w^{(j)}_{,\zeta\zeta}, && \bar{h} = \frac{D_{22}}{D_{66}} \left(w^{(j)}_{,\chi\chi\chi} + \frac{D_{12}}{D_{22}} w^{(j)}_{,\zeta\zeta\chi} \right) + 4 w^{(j)}_{,\zeta\zeta\chi},
\end{aligned}
\tag{3.37}
$$

where

$$
\zeta = \frac{x}{b},\ \chi = \frac{y}{b} \tag{3.38}
$$

The functions $\bar{a}$, $\bar{b}$, $\bar{g}$, $\bar{h}$ are related to sectional generalized forces (3.14) and (3.15) for the first order approximation.

After taking into account relations (3.37) into (3.36), the following homogeneous system of differential equations is obtained:

$$
\begin{cases}
\bar{a}_{,\chi} = -\frac{A_{66}}{A_{22}} \bar{b}_{,\zeta} + \lambda \Delta \frac{A_{11}}{A_{22}} \left(1 - \frac{A_{12}}{A_{11}} \frac{A_{12}}{A_{22}} \right) \bar{d}_{,\zeta\zeta} + h\rho \ddot{\bar{d}},\\
\bar{b}_{,\chi} = -\frac{A_{11}}{A_{66}} \left(\bar{c}_{,\zeta\zeta} + \frac{A_{12}}{A_{11}} \bar{d}_{,\zeta\chi} \right) + h\rho \frac{A_{11}}{A_{66}} \ddot{\bar{c}},\\
\bar{c}_{,\chi} = \bar{b} - \bar{d}_{,\zeta},\\
\bar{d}_{,\chi} = \bar{a} - \frac{A_{12}}{A_{22}} \bar{c}_{,\zeta},\\
\bar{e}_{,\chi} = \bar{f},\\
\bar{f}_{,\chi} = \bar{g} - \frac{D_{12}}{D_{22}} \bar{e}_{,\zeta\zeta},\\
\bar{g}_{,\chi} = \frac{D_{66}}{D_{22}} \left(\bar{h} - 4 \bar{f}_{,\zeta\zeta} \right),\\
\bar{h}_{,\chi} = \frac{A_{11}}{D_{66}} \left[-\lambda \Delta \left(1 - \frac{A_{12}}{A_{11}} \frac{A_{12}}{A_{22}} \right) \bar{e}_{,\zeta\zeta} - \frac{D_{11}}{A_{11}} \bar{e}_{,\zeta\zeta\zeta\zeta} - \frac{D_{12}}{A_{11}} \bar{f}_{,\zeta\zeta\chi} \right] + h\rho \frac{A_{11}}{D_{66}} \ddot{\bar{e}}.
\end{cases}
\tag{3.39}
$$

The solution to (3.39) consistent with boundary conditions (3.2) has been adopted in the form of the following series of functions:

$$\begin{aligned}
\overline{a}_i &= T_m(t)\overline{A}_i(\chi_i)\sin\tfrac{m\pi b_i}{l}\zeta_i, & \overline{b}_i &= T_m(t)\overline{B}_i(\chi_i)\cos\tfrac{m\pi b_i}{l}\zeta_i,\\
\overline{c}_i &= T_m(t)\overline{C}_i(\chi_i)\cos\tfrac{m\pi b_i}{l}\zeta_i, & \overline{d}_i &= T_m(t)\overline{D}_i(\chi_i)\sin\tfrac{m\pi b_i}{l}\zeta_i,\\
\overline{e}_i &= T_m(t)\overline{E}_i(\chi_i)\sin\tfrac{m\pi b_i}{l}\zeta_i, & \overline{f}_i &= T_m(t)\overline{F}_i(\chi_i)\sin\tfrac{m\pi b_i}{l}\zeta_i,\\
\overline{g}_i &= T_m(t)\overline{G}_i(\chi_i)\sin\tfrac{m\pi b_i}{l}\zeta_i, & \overline{h}_i &= T_m(t)\overline{H}_i(\chi_i)\sin\tfrac{m\pi b_i}{l}\zeta_i,
\end{aligned} \tag{3.40}$$

where $T_m(t)$ is an unknown function of time, and the functions $\overline{A}_i, \overline{B}_i, \overline{C}_i, \overline{D}_i, \overline{E}_i, \overline{F}_i, \overline{G}_i, \overline{H}_i$ are unknown functions of y in the transverse direction. These functions are determined during the numerical calculations in the first order approximation by the modified numerical transition matrix method, taking into consideration the Godunov orthogonalization method [1, 18, 19], which assumes that the functions correspond to the modes of undamped natural frequencies.

An adoption of expected solution (3.40) as a function of time:

$$T_m(t) = e^{j\omega_m t}, \tag{3.41}$$

where $j = \sqrt{-1}$ and ω_m are natural frequencies for an m number of halfwaves in the longitudinal direction, allows for a modal analysis including an influence of external load on the frequency of vibration [40].

The further solution is identical to the buckling problem solution described in Sect. 3.1.

It should be noted that many authors in the world literature (for example, [41]) have not taken into account inertia forces in the plane of the wall (i.e., $h\rho\,\ddot{u}$ and $h\rho\,\ddot{v}$) in differential equilibrium Eq. (3.35). Such an approach may lead to differences in results.

Differences in the natural frequency corresponding to the flexural buckling mode with a single halfwave ($m = 1$) change with the length of the column (Fig. 3.3). For relatively long columns (length to width of the widest wall $l/b_{max} \geq 15$), these differences reach even more than 42 % (Fig. 3.4). Thus, it should be said that neglecting tangential inertia leads to considerable errors for long thin-walled columns. However, the natural frequencies corresponding to local buckling modes show almost no differences (a percentage difference is about 0.1 %). Therefore, in the analysis of the natural frequency corresponding to the local mode, the inertial force acting in the plane of the plate or the wall of the column can be omitted [31].

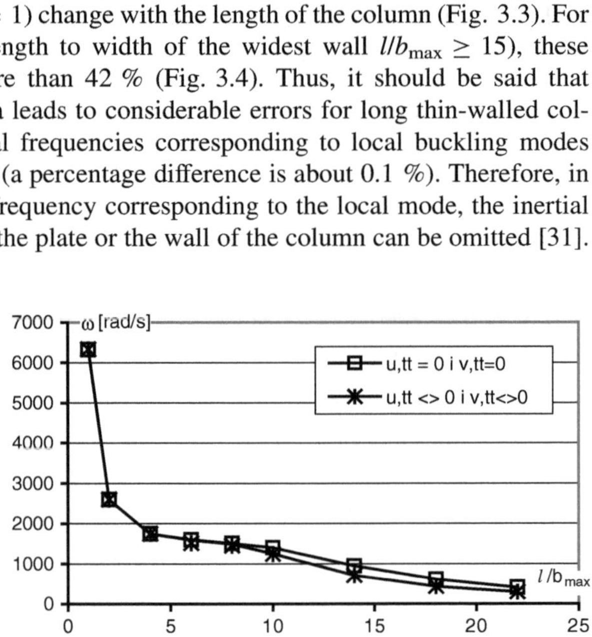

Fig. 3.3 Natural frequencies of orthotropic columns of different length (dimensions to be found in Fig. 7.1, material properties—in Table 6.1)

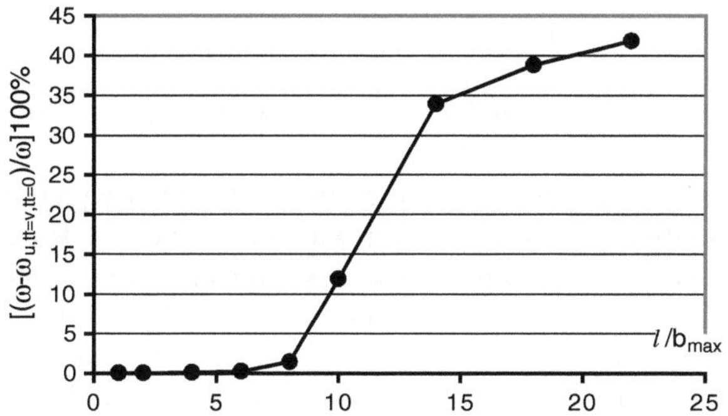

Fig. 3.4 Percentage differences in the natural frequency for a single halfwave mode

3.3 Dynamic Buckling

In the dynamic analysis (while finding the frequency of natural vibrations [40]), the independent nondimensional displacement ξ and the load factor λ become functions dependent on time, and dynamic terms are added to the equations describing the postbuckling equilibrium path. Neglecting the forces associated with the inertia terms of the prebuckling state and the second order approximation and taking into account the orthogonality conditions for the displacement field in the first $\bar{U}^{(j)}$and second order approximation $\bar{U}^{(ij)}$, the Lagrange equations can be written as [28]:

$$\frac{1}{\omega_s^2}\ddot{\xi}_s+\left(1-\frac{\lambda}{\lambda_s}\right)\xi_s+a_{ijs}\xi_i\xi_j+b_{ssss}\xi_s^3=\xi_s^*\frac{\lambda}{\lambda_s};\ (s=1,2,\ldots,N), \tag{3.42}$$

where ω_s is a natural frequency with the mode corresponding to the buckling mode; a_{ijs} and b_{ssss} are coefficients (3.31) describing the postbuckling behaviour of the structure (independent of time); however, parameters of the load λ and the displacement ξ are functions of time t.

For the uncoupled buckling, i.e., the single-mode buckling (where index $s = N = 1$), the equations of motion may be written in the form:

$$\frac{1}{\omega_1^2}\ddot{\xi}_1+\left(1-\frac{\lambda}{\lambda_1}\right)\xi_1+a_{111}\xi_1^2+b_{1111}\xi_1^3=\xi_1^*\frac{\lambda}{\lambda_1}. \tag{3.43}$$

It is assumed that in the initial moment of time $t = 0$, the nondimensional displacements ξ, as well as displacement velocities, are equal to zero, i.e.:

$$\xi(t=0) = 0\text{and } \dot{\xi}(t=0)=0. \tag{3.44}$$

The Runge-Kutta method [34] for solving (3.43) requires the following substitutions:

$$\begin{aligned} \dot{\xi} &= \Gamma(t), \\ \dot{\Gamma} &= -\omega_1^2\left(1-\frac{\lambda(t)}{\lambda_1}\right)\xi - \omega_1^2 b_{111}\xi_1^2 - \omega_1^2 b_{1111}\xi_1^3 + \omega_1^2\frac{\lambda(t)}{\lambda_1}\xi^*, \end{aligned} \tag{3.45}$$

which lead to a system of two differential equations. ''Complete'' equations of motion (3.45) are solved with the numerical Runge-Kutta method. In order to obtain a result with a fixed precision and speed up the computational algorithm, the Dormad-Prince eighth order method [34] with the Hairer and Wanner [3] modification for determining the integration step has been employed. In the procedure proposed by Hairer and Wanner, the fifth-order Runge–Kutta method was used and to assess the accuracy—the third order method.

References

1. Biderman VL (1977) Mekhanika tonkostennykh konstruktsii. Mashinostroenie, Moscow
2. Byskov E, Hutchinson JW (1977) Mode interaction in axially stiffened cylindrical shells. AIAA J 15(7):941–948
3. Hairer E, Nørsett S, Wanner G (1993) Solving ordinary differential equations. Springer Verlag, New York
4. Koiter WT (1963) Elastic stability and post-buckling behaviour. In: Proceedings of the symposium on non-linear problems, University of Wisconsin Press, Wisconsin, pp 257–275
5. Koiter WT (1976) General theory of mode interaction in stiffened plates and shell structures. WTHD Report 590, Delft University, Netherland
6. Kolakowski Z (1987) Mode interaction in thin-walled trapezoidal column under uniform compression. Thin Wall Struct 5:329–342
7. Kolakowski Z (1993) Interactive buckling of thin-walled column-beams with open and closed cross-sections. Thin Wall Struct 15(3):159–183
8. Kolakowski Z (1993) Influence of modification of boundary conditions on load carrying capacity in thin-walled columns in the second order approximation. Int J Solids Struct 30(19):2597–2609
9. Kolakowski Z (2007) Some aspects of dynamic interactive buckling of composite columns. Thin Wall Struct 45(10–11):866–871
10. Kolakowski Z, Kowal-Michalska K (eds) (1999) Selected problems of instabilities in composite structures. A series of monographes. Technical University of Lodz Press, Lodz
11. Kolakowski Z, Kowal-Michalska K (eds) (2012) Static, dynamic and stability of structures, vol 2, statics, dynamics and stability of structural elements and systems, a series of monographs. Technical University of Lodz Press, Lodz
12. Kolakowski Z, Kowal-Michalska K (2012) Lower bound estimation of post-buckling behaviour of thin-walled functionally graded structures. In: Ferreira AJM, Carrera E (eds) Proceedings of mechanics of nano, micro and macro composite structures, Politecnico di Torino, 18–20 June 2012
13. Kolakowski Z, Krolak M (1995) Interactive elastic buckling of thin-walled closed orthotropic beam-columns. Eng Trans 43(4):591–602
14. Kolakowski Z, Krolak M (2006) Modal coupled instabilities of thin-walled composite plate and shell structures. Compos Struct 76:303–313
15. Kolakowski Z, Kubiak T (2004) Multiple interaction of dynamic buckling modes in thin-walled members subjected in-plane pulse loading. In: Proceedings of 4th international conference on coupled instabilities in metal structure, Rome, 27–29 Sept 2004

16. Kolakowski Z, Kubiak T (2007) Interactive dynamic buckling of orthotropic thin-walled channels subjected to in-plane pulse loading. Compos Struct 81:222–232
17. Kolakowski Z, Mania RJ (2013) Semi-analytical method versus the FEM for analysis of the local post-buckling of thin-walled composite structures. Compos Struct 97:99–106
18. Kolakowski Z, Teter A (1995) Interactive buckling of thin-walled closed elastic beam-columns with intermediate stiffeners. Int J Solids Struct 32(11):1501–1516
19. Kolakowski Z, Teter A (2000) Interactive buckling of thin-walled beam-columns with intermediate stiffeners or/and variable thickness. Int J Solids Struct 37(24):3323–3344
20. Kolakowski Z, Krolak M, Kowal-Michalska K (1999) Modal interactive buckling of thin-walled composite beam-columns regarding distortional deformations. Int J Eng Sci 37:1577–1596
21. Krolak M (ed) (1990) Stany zakrytyczne i nosnosc graniczna cienkosciennych dzwigarow o scianach plaskich. PWN, Warsaw—Lodz
22. Krolak M (ed) (1995) Statecznosc, stany zakrytyczne i nosnosc graniczna cienkosciennych dzwigarow o scianach plaskich. Series of monographs. Technical University of Lodz Press, Lodz
23. Krolak M, Kolakowski Z (1995) Interactive elastic buckling of thin-walled open orthotropic beam-columns. Eng Trans 43(4):571–590
24. Krolak M, Mania RJ (eds) (2011) Static, dynamic and stability of structures, vol 1, stability of thin-walled plate structures. A series of monographs. Technical University of Lodz Press, Lodz
25. Kubiak T (1998) Nieliniowa analiza statecznosci ortotropowych cienkosciennych pretow o roznych ksztaltach przekrojow poprzecznych, Ph.D. thesis, Technical University of Lodz, Lodz
26. Kubiak T (2001) Postbuckling behavior of thin-walled girders with orthotropy varying widthwise. Int. J. Solid Struct 38(28–29):4839–4856
27. Kubiak T (2007) Interakcyjne wyboczenie dynamiczne cienkosciennych slupow. Technical University of Lodz Press, Lodz
28. Kubiak T (2011) Estimation of dynamic buckling for composite columns with open cross-section. Comput Struct 89(21–22):2001–2009
29. Manevich AI, Kolakowski Z (1996) Influence of local postbuckling behaviour on bending of thin-walled beams. Thin Wall Struct 25(3):219–230
30. Moellmann M, Goltermann P (1989) Interactive buckling in thin-walled beams, Part I: theory, Part II: applications. Int J Solids Struct 25(7):715–728
31. Pietraszkiewicz W (1989) Geometrically nonlinear theories of thin elastic shells. Adv Mech 12(1):51–130
32. Pignataro M, Luongo A (1987) Asymmetric interactive buckling of thin-walled columns with initial imperfection. Thin Wall Struct 3:365–385
33. Pignataro M, Luongo A, Rizzi N (1985) On the effect of the local overall interaction on the post-buckling of uniformly compressed channel. Thin Wall Struct 3:283–321
34. Prince PJ, Dormand JR (1981) High order embedded Runge-Kutta Formulae. J Comp Appl Math 7:67–75
35. Sridharan S, Ali MA (1986) Interactive buckling in thin-walled beam-columns. J Eng Mech ASCE 111(12):1470–1486
36. Sridharan S, Ali MA (1986) An improved interactive buckling analysis of thin-walled columns having doubly symmetric sections. Int J Solids Struct 22(4):429–443
37. Sridharan S, Benito R (1984) Columns static and dynamic interactive buckling. J Eng Mech, ASCE 110(1):49–65
38. Sridharan S, Graves-Smith TR (1981) Post-buckling analyses with finite strips. J Eng Mech, ASCE 107(EM5):869–888
39. Sridharan S, Peng MH (1989) Performance of axially compressed stiffened panels. Int J Solids Struct 25(8):879–899
40. Teter A, Kolakowski Z (2003) Natural frequencies of a thin-walled structures with central intermediate stiffeners or/and variable thickness. Thin Wall Struct 41:291–316
41. Wozniak Cz (ed) (2001) Mechanika sprezystych plyt i powlok. PWN, Warsaw

Chapter 4
Finite Element Method

The finite element method is used here to validate the analytical–numerical method presented in Chap. 3. The basic assumption and the used equations for thin plates in both the methods (ANM—the analytical–numerical method, FEM—the finite element method) are identical, but the way of solution is different. To apply a program based on the finite element method as a numerical experiment, confirming or validating the theoretical analysis or the analysis based on the analytical–numerical method, it is important to create an appropriate model. A properly chosen finite element type, a rational mesh density and appropriate boundary conditions play a significant role in obtaining the calculation results close to the reality.

In the proposed analytical–numerical method, the solution is based on the assumed deflection in the form of the sine function (3.21) in the longitudinal direction and is calculated using the numerical transition matrix method in the transverse direction. In the finite element method, the deflection function is called a shape function and it is usually assumed for each element as the first or second order polynomial. The final deflection of the whole structure is the effect of displacement of each element.

For the analysed thin-walled structures, finite shell elements seem to be the best choice. Their size and number (mesh) have to be chosen so as to map smoothly deformations. Assuming the number of elements for a dynamic buckling problem analysis, one should be aware that the map of deflection could be different than for static buckling—a number of sine halfwaves in the longitudinal direction could be greater than in the case of static load.

It seems that an application of the finite element method to replace the experiment is an easier and much cheaper option. However, please note that boundary conditions have a significant impact on the results obtained and must be the same (or similar) in all the test methods compared.

The finite element method becomes more and more popular and it represents one of the most significant developments in the history of computational methods. The finite element method has transformed much of theoretical mechanics and

T. Kubiak, *Static and Dynamic Buckling of Thin-Walled Plate Structures*,
DOI: 10.1007/978-3-319-00654-3_4,

abstract science into practical and essential tools for a multitude of technological developments which effect many facets of our life.

It is difficult to document the exact origin of the FEM but it can be said that the basic concept has evolved over a period of one hundred years or more.

The FEM theory started in the 1940s. The first formulations were developed as matrix methods for structural analysis. This led to the idea to approximate solids and Courant [4] introduced it as an assembly of triangular elements and the minimum of potential energy to torsion problems. Shortly thereafter, Clough [26] introduced the term "finite element" in the paper published together with Turner, Martin and Topp. Their paper focused on the "stiffness and deflection of complex structures". The finite element method was further enhanced during the 1960s and 1970s by such scientists as Zienkiewicz [28, 30], Hinton and Owen [10]. Zienkiewicz and Cheung [29] applied the technology to general problems described by Laplace and Poisson's equations. The major contributor among mathematicians who were developing better solution algorithms and carried out the modelling and solution of nonlinear problems was Crisfield [5]. Starting since the 1970s also Polish scientists have been concerned with the FEM—the leader was Szmelter [23, 24]. In the 1990s a rapid increase in the computing power contributed to a sudden expansion and propagated the finite element method. This method was applied to different software, which can be used in many disciplines.

Nowadays the finite element analysis is used not only for solving engineering problems but it is also used by scientists as a numerical experiment. By introducing new elements and mathematical techniques, the method has been still developing.

4.1 Dynamic Buckling

To solve the dynamic buckling problem, a response of the structure subjected to pulse load should be known. To describe the above-mentioned behaviour, the equations of motion should be solved. In the finite element method, exactly the same as in the analytical-numerical method, the differential equations of motion of the plate are derived on the basis of the Hamilton's principle (2.48):

$$\delta\Psi = \delta\int_{t_0}^{t_1}\Lambda dt = \delta\int_{t_0}^{t_1}(K-\Pi)dt = 0, \tag{4.1}$$

where K is a kinetic energy of the system and Π is a total potential energy of the system.

Similarly to the derivation presented in Sect. 2.5, the kinetic energy is:

$$K = \frac{1}{2}\rho\int_{\Omega}\{\dot{U}\}^2 d\Omega, \tag{4.2}$$

and its variation, taking into consideration identity (2.60), can be written as follows:

$$\delta K = \int_{\Omega} \rho\{\delta U\}^T\{\dot{U}\}d\Omega, \tag{4.3}$$

where {U} is a vector of the displacement function which contains the following elements: u(x,y,z,t), v(x,y,z,t), w(x,y,z,t)—functions describing displacements of a given point in three perpendicular directions in the given moment of time.

Integrating the kinetic energy variation over time, the following relation is obtained:

$$\begin{aligned}\int_{t_0}^{t_1} \delta K dt &= \int_{t_0}^{t_1}\int_{\Omega} \rho\{\delta U\}^T\{\dot{U}\}d\Omega dt \\ &= \int_{\Omega} \rho\{\delta U\}^T\{\dot{U}\}d\Omega\Big|_{t_1}^{t_2} d\Omega - \int_{t_0}^{t_1}\int_{\Omega} \rho\{\delta U\}^T\{\ddot{U}\}d\Omega dt,\end{aligned} \tag{4.4}$$

where the first term vanishes because the displacement variation $\{\delta \mathrm{U}\}$ equals zero for $t = t_0$ and $t = t_1$.

The total potential energy variation $\delta\Pi$ can be written in the form:

$$\delta\Pi = \delta Q - \delta W, \tag{4.5}$$

where δQ is a variation of the internal elastic strain energy:

$$\delta Q = \int_{\Omega} \{\delta\varepsilon\}^T\{\sigma\}d\Omega, \tag{4.6}$$

and δW is a variation of work of the external forces $\{F\}$:

$$\delta W = \int_{\Omega} \{\delta U\}\{F\}d\Omega. \tag{4.7}$$

Substituting (4.4) and (4.5) into (4.1), the Hamilton's principle can be written in the form:

$$\int_{t_0}^{t_1}\int_{\Omega} [\rho\{\delta U\}^T\{\ddot{U}\} - \{\delta\varepsilon\}^T\{\sigma\} + \{\delta U\}^T\{F\}]d\Omega dt = 0. \tag{4.8}$$

In the finite element method, the displacement {U} for any point of the structure should be related to the nodal displacements {u} using an arbitrary assumed shape function [$\mathrm{N_u}$]:

$$\{U\} = [N_u]\{u\}. \tag{4.9}$$

The shape function of the element applied to discretize the structure is also used to express:

- strains $\{\varepsilon\}$ by the vector of nodal displacements {u} and derivatives of the shape function called the geometrical matrix $[B_{\mathrm{u}}]$:

$$\{\varepsilon\} = [D][N_u]\{u\} = [B_u]\{u\}, \tag{4.10}$$

where:

$$[D] = \begin{bmatrix} \partial_x & 0 & 0 \\ 0 & \partial_y & 0 \\ 0 & 0 & \partial_z \\ 0 & \partial_z & \partial_y \\ \partial_z & 0 & \partial_x \\ \partial_y & \partial_x & 0 \end{bmatrix}, \ \partial_x = \frac{\partial}{\partial x}, \ \partial_y = \frac{\partial}{\partial y}, \ \partial_z = \frac{\partial}{\partial z}, \tag{4.11}$$

- stress $\{\sigma\}$ by the vector of nodal displacements {u}, the geometrical matrix $[B_{\mathrm{u}}]$ and the material properties matrix [Q]:

$$\{\sigma\} = [Q]\{\varepsilon\} = [Q][B_u]\{u\}. \tag{4.12}$$

Substituting (4.9)–(4.12) to the Hamilton's principle, i.e., to (4.8), the following form is obtained:

$$\{\delta u\}^T \int\limits_{\Omega} [\rho[N_u]^T[N_u]d\Omega\{\ddot{u}\} - \{\delta u\}^T \int\limits_{\Omega} [B]^T[Q][B]d\Omega\{u\} + \{\delta u\}^T \int\limits_{\Omega} [N_u]^T\{F\}d\Omega = 0, \tag{4.13}$$

where the integration over time is omitted for simplicity. The integration requires a numerical procedure which is carried out in discretized moments with the established time step of integration. The Newmark method [1, 2] is a very popular procedure also used in the ANSYS software.

Equation (4.13) must be satisfied for any variation of displacements (weight functions) in any moment of time from t_0 to t_1 and fulfil the relevant boundary conditions. This is a weak (variation) formulation of finite element method equations, where the generalized Galerkin method [3, 21, 27] has been used. Equating to zero the coefficients standing by the variations of displacements, the following equation is obtained in the matrix form:

$$\{P\} = [M] \cdot \{\ddot{u}\} + [K] \cdot \{u\}, \tag{4.14}$$

where:

$[M] = \int\limits_{\Omega} [\rho[N_u]^T [N_u] d\Omega$ is a mass matrix of the structure,

$[K] = \int\limits_{\Omega} [B]^T [Q][B] d\Omega$ is a structural stiffness matrix,

$\{P\} = \int\limits_{\Omega} [N_u]^T \{F\} d\Omega$ is a vector of generalized nodal forces.

After replacing the time derivatives of displacements $\{\ddot{u}\}$ by the displacement differences $\{u\}$ in successive discrete moments of time t, a new static equilibrium equation for the each time step is obtained. It contains the inertia forces [M] $\{\ddot{u}\}$ and, therefore, it is possible to apply the algorithms used in the static analysis. The time integration in the ANSYS program is conducted using the Newmark method [2] and a solution to equations in successive time steps is obtained with the Newton-Raphson algorithm [7].

Equation (4.14) does not take into account the material-dependent damping, which according to [13] can be neglected in the dynamic buckling analysis of thin-walled structures subjected to uniform compression. However, the recent paper [14] has shown a significant role of dumping in the case when the viscoelastic material model with the strain rate effect is taken into account and/or the analysed structures are subjected to torsion or combined load (bending and torsion). In this case, the damping related to the speed of displacement should be added to (4.14), which takes the form:

$$\{P\}=[M] \times \{\ddot{u}\}+[C] \times \{\dot{u}\}+[K] \times \{u\}, \tag{4.15}$$

where [C] is a damping matrix, which according to the ANSYS manual [1] depends on the mass matrix of structures [M] and the structural stiffness matrix [K]. The relation is as follows:

$$[C] = \alpha[M] + \beta[K], \tag{4.16}$$

where α and β are damping coefficients specified as decimal numbers. The values of α and β [1] are not generally known directly, but can be calculated from the modal damping ratio ψ_i, which is the ratio of actual damping to critical damping for the i-th mode of vibration. Denoting the natural circular frequency of the i-th mode by ω_i, the coefficients α and β satisfy the relation:

$$\psi_i = \frac{\alpha}{2\omega_i} + \frac{\beta\omega_i}{2}. \tag{4.17}$$

In many practical structural problems, the alpha damping (or the *mass* damping) may be ignored ($\alpha = 0$). In such cases, the beta damping β can be evaluated from known values of ψ_i and ω_i as:

$$\beta = \frac{2\psi_i}{\omega_i}, \tag{4.18}$$

and then the damping matrix can be expressed as follows:

$$[C] = \beta[K]. \tag{4.19}$$

It should be noted that (in the ANSYS software [1]):

- in the transient harmonic response analysis, the *material-dependent* damping can be only specified as the beta damping (β), using the command defining material properties;
- only one β can be input in each load step, so the most dominant frequency ω_i should be chosen to calculate the beta damping coefficient.

The dynamic buckling analysis or the response analysis of the structure subjected to pulse load requires the following assumptions:

- shape and amplitude of imperfection,
- shape and amplitude (defined as DLF) of pulse load,
- time of pulse duration.

Therefore, before attempting to solve the dynamic buckling problem, the eigenbuckling solution (Sect. 4.3) should be achieved to determine the critical load and the corresponding buckling mode. The buckling mode corresponding to the lowest critical load is used to map the shape of geometrical imperfections with the assumed amplitude, usually equal to 1/100 thickness of the plate or the wall of the column. The modal analysis (Sect. 4.4) is performed to calculate the natural frequencies of the structure which are used to determine the duration of pulse load. The pulse duration T_p is usually assumed as a half or one period of the natural vibration T.

The above discussed analysis has been carried out using the ANSYS software [1] based on the finite element method.

4.2 Nonlinear Buckling Analysis

A nonlinear buckling analysis is in fact a static analysis with geometrical nonlinearities (large deflections included into the analysis). This kind of analysis allows one to find the buckling load, to analyse the postbuckling behaviour of structures, to estimate the load carrying capacity and a mode of failure.

For the structures which are subject to perfectly in-plane load (uniformly compressed plates or columns), small out-of-plane perturbations are necessary to initiate the buckling response. These perturbations can be applied as a modest temporary force or a specified displacement. It should be noted that the failure load is very sensitive to initial imperfections, therefore, these initial perturbations should be as small as possible. Usually, at the beginning the eigenbuckling analysis is performed to calculate the buckling load and the corresponding buckling mode. The shape of the buckling mode allows one to predict the place where the force as the initial perturbation should be applied. In the case of the initial displacement, the buckling mode is used as the perturbation to map it into the finite element model.

In other cases of load, i.e., out-of-plane load or in-plane load but with eccentricity, the initial perturbation is not necessary.

Having the proper model, the following equation is solved:

$$[\overline{\mathrm{K}}] \cdot \{\mathrm{u}\} = \{\mathrm{P}\}, \tag{4.20}$$

where the stiffness matrix $[\overline{\mathrm{K}}]$ is dependent on the nodal displacement, which means that the static problem is no longer geometrically linear—it is a nonlinear problem (in the ANSYS software, the nonlinear geometry procedure should be switched on). From the theoretical point of view, this nonlinear static problem can be treated as a large strain or large rotation problem. In the first case (i.e., large strain), strains in the material exceed more than a few percent—it usually happens when a hyperelastic problem or an elastic-plastic problem is solved. In the second case (large rotation), rotations are large and mechanical strains are small. The second case describes situations similar to the postbuckling analysis in an elastic range, especially when the deflections of thin-walled structures are less than 5 thicknesses of the plate or girder wall—the results are similar to the one obtained from the analytical-numerical method described in Chap. 3.

The stiffness matrix $[\overline{\mathrm{K}}]$ should be treated as the tangent stiffness matrix for the i-th increment of displacement and it is usually denoted as $[K_i^T]$. Depending on the type of geometric nonlinearity, the tangent stiffness matrix $[K_i^T]$ and the vector of restoring forces for the i-th iteration $\{P_i^r\}$ are formulated in a different way. According to the notation used in (4.20), the vector of restoring force (at each iteration) is a part of the vector {P}, which can be expressed as:

$$\{P\} = \{P^a\} - \{P_i^r\}, \tag{4.21}$$

where $\{P^{\mathrm{a}}\}$ is a vector of applied nodal forces.

For a large strain problem formulation, the tangent stiffness matrix is as follows:

$$[K_i^T] = [K_i] + \int_{\Omega} [G_i]^T [\tau_i][G_i] d\Omega, \tag{4.22}$$

where $[\mathrm{K_i}]$ is an elemental stiffness matrix (well known from the static linear problem formulation), $[\mathrm{G_i}]$ is a matrix of shape function derivatives and $[\tau_\mathrm{i}]$ is a matrix of the current Cauchy (true) stresses $\{\sigma_\mathrm{i}\}$.

The vector of restoring forces corresponding to the element internal loads for the i-th iteration $\{P_i^r\}$ in the large strain problem formulation is:

$$\{P_i^r\} = \int_\Omega [B_i]\{\sigma_i\}\mathrm{d}\Omega. \tag{4.23}$$

For the large rotation problem formulation, the tangent stiffness matrix is as follows:

$$[K_i^T] = \int_\Omega [T_n]^T[B_V]^T[D][B_V][T_n]d\Omega, \tag{4.24a}$$

where $[B_v]$ is a small strain-displacement relationship in the original (virgin—non-deformed) element coordinate system and $[T_n]$ is an orthogonal transformation relating the original element coordinates to the converted (or rotated) element coordinates. The rotated element coordinate system differs from the original coordinate system by the amount of rigid body rotation. Hence, the transformation matrix $[T_n]$ is computed by separating the rigid body rotation from the total deformation $\{u_n\}$ using the polar decomposition theorem (the deformation of the object could be decomposed as the rotation plus stretching).

The vector of restoring forces for the *i*-th iteration $\{P_i^r\}$ in the large rotation problem formulation is:

$$\{P_i^r\} = \int_\Omega [T_n]^T[B_V]^T[D][B_V]\{u_d\}d\Omega, \tag{4.24b}$$

where $\{u_d\}$ is the element deformation which causes straining [1].

The Newton-Raphson method, which is graphically presented in Fig. 4.1, is most popular for solving nonlinear problems with the iterative method. For a single iteration, (4.20) has the form [2]:

$$[K_i^T]\{\Delta u_i\} = \{P^a\} - \{P_i^r\}, \tag{4.25}$$

where the increment of displacement $\{\Delta u_i\}$ is assumed and fulfils the following relation:

$$\{u_{i+1}\} = \{u_i\} + \{\Delta u_i\}. \tag{4.26}$$

The general algorithm for the Newton-Raphson procedure (Fig. 4.1) can be described as the general algorithm and proceeds as follows:

1. Assume an initial displacement vector $\{u_0\}$. In the first time step, the initial displacement vector is assumed as zero $\{u_0\} = \{0\}$ and for next time steps, $\{u_0\}$ is assumed as the displacement vector from the previous time step converged solution.
2. Compute the updated tangent matrix $[K_i^T]$ and the restoring load vector $\{P_i^r\}$ for the displacement vector $\{u_i\}$ for the *i*-th iteration.
3. Calculate a vector of the displacement increment $\{\Delta u_i\}$ from (4.25).
4. Find the next approximation $\{u_{i+1}\}$ from (4.26).

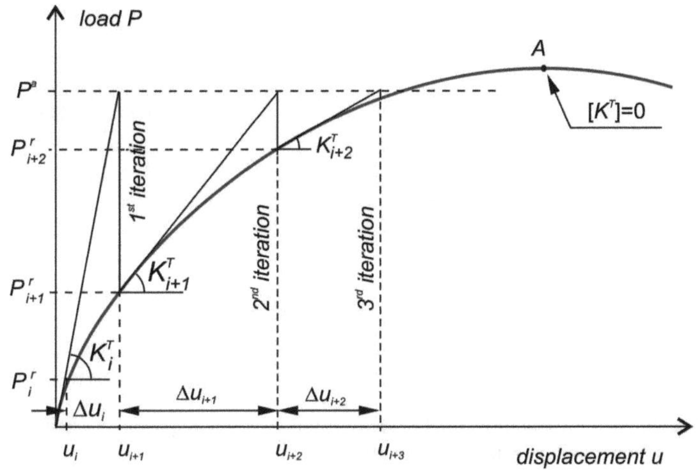

Fig. 4.1 Graphical representation of the Newton-Raphson algorithm

5. Repeat steps 2–4 until a convergence is obtained.

In practice, more than one iteration is needed to obtain a converged solution. In order to improve and speed up the procedure, the incremental Newton-Raphson procedure (Fig. 4.2a) or the initial-stiffness Newton-Raphson procedure (Fig. 4.2b) is used. In the incremental procedure, the applied load is divided in a few substeps (Fig. 4.2a). Using the initial-stiffness Newton-Raphson procedure, the tangent stiffness matrices are the same during each iteration, which requires fewer matrix reformulations and inversions than in the full Newton-Raphson procedure. It should be also mentioned that due to numerous iterations, the initial-stiffness Newton-Raphson procedure converges more slowly than the full Newton-Raphson procedure.

The main disadvantage of the Newton-Raphson method is such that it stops the procedure when the determinant of the tangent stiffness matrix is equal to zero (point A in Fig. 4.1).

Therefore, the arc-length technique (Riks method) is used. The Riks method [22] is suitable for nonlinear static equilibrium solutions to unstable problems, allows for finding the load carrying capacity and for analysing a failure mode. The arc-length method in the ANSYS software uses the explicit spherical iterations to maintain the orthogonality between the arc-length radius and orthogonal directions as described by Forde and Stiemer [6]. A graphical representation is shown in Fig. 4.3. It is assumed that all load magnitudes in (4.25) are controlled by the total load factor λ, which is in the range $\langle -1,1 \rangle$. Then, in the arc-length method, (4.25) has the form:

$$\left[K_i^T\right]\{\Delta u_i\} = \lambda\{P^a\} - \left\{P_i^r\right\}. \tag{4.27}$$

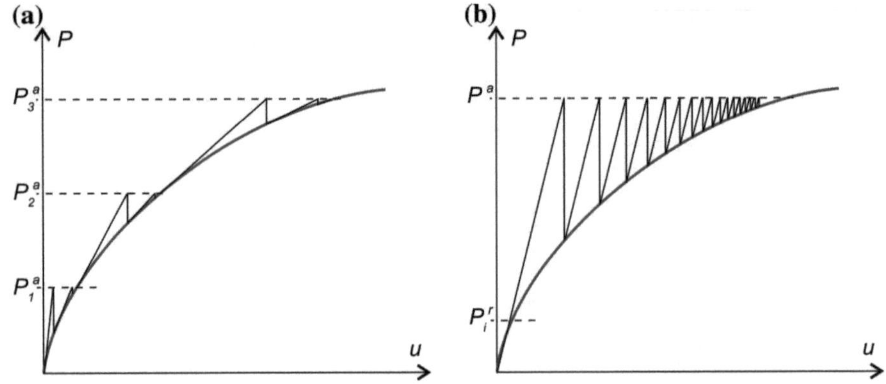

Fig. 4.2 Graphical representation of the incremental Newton-Raphson procedure (**a**) and the initial-stiffness Newton-Raphson procedure (**b**)

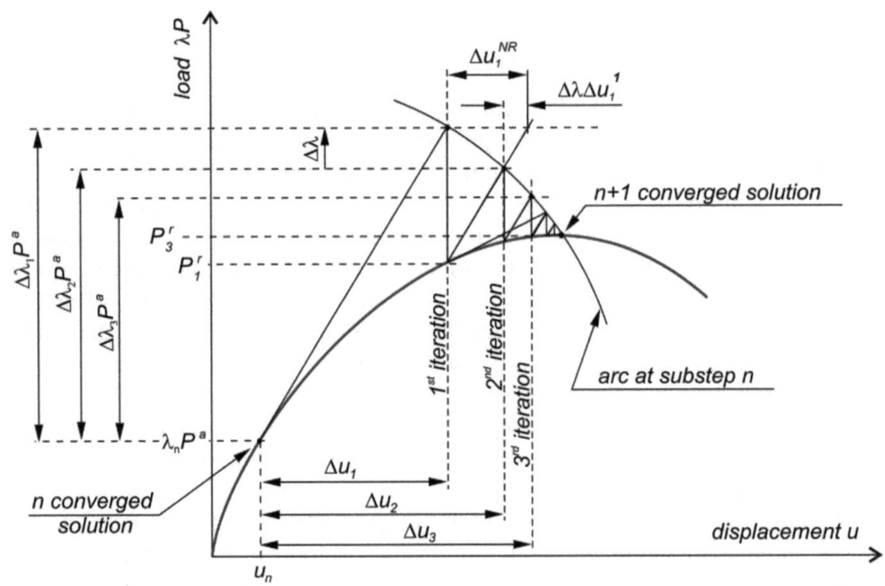

Fig. 4.3 Graphical representation of the arc-length method

Equation (4.27) in the incremental form for the intermediary step (i.e., at the substep n and the iteration i) has the following form:

$$\left[K_i^T\right]\{\Delta u_i\} - \Delta\lambda\{P^a\} = (\lambda_n + \lambda_i)\{P^a\} - \left\{P_i^r\right\}, \tag{4.28}$$

where $\Delta\lambda$ is an incremental load factor (Fig. 4.3).

On the basis of (4.28), the relation describing the vector of incremental displacement $\{\Delta u_i\}$ can be written as follows:

$$\{\Delta u_i\} = \Delta\lambda\{\Delta u_i^1\} + \{\Delta u_i^{NR}\}, \tag{4.29}$$

where $\{\Delta u_i^1\}$ is the vector of incremental displacement corresponding to a unit load factor and $\{\Delta u_i^{NR}\}$ is the vector of incremental displacement taken from the conventional Newton-Raphson method (4.25). Both the vectors of incremental displacement are defined by:

$$\{\Delta u_i^1\} = [K_i^T]^{-1}\{P^a\}, \tag{4.30}$$

$$\{\Delta u_i^{NR}\} = [K_i^T]^{-1}((\lambda_n + \lambda_i)\{P^a\} - \{P_i^n\}), \tag{4.31}$$

and both of them should be determined in each arc-length iteration. Then, the incremental load factor $\Delta\lambda$ is determined by the arc-length l_i equation, which at the i-th iteration can be written as (Fig. 4.3):

$$l_i^2 = \Delta\lambda_i^2 + \beta^2\{\Delta u_n\}^T\{\Delta u_n\}, \tag{4.32}$$

where β is a scaling factor, Δu_n is a sum of all the displacement increments Δu_i of this iteration.

Finally, the arc-length load increment factor $\Delta\lambda$ is determined by the formula:

$$\Delta\lambda = \frac{r_i - \{\Delta u_n\}^T\{\Delta u_i^{NR}\}}{\beta^2\lambda_i + \{\Delta u_n\}^T\{\Delta u_i^1\}}, \tag{4.33}$$

where r_i is a residual parameter (a scalar) obtained by multiplication of the normal and tangential vectors.

The arc-length method has also disadvantages and the main one is the requirement to adjust the arc-length radius by trial-and-error in a series of manually directed reanalyses.

4.3 Linear Static Buckling Analysis

The linear buckling analysis of the structure subjected to static load is called the eigenbuckling problem. The eigenbuckling is used for bifurcation buckling of the linearized model of elastic stability. This type of analysis allows for calculating the buckling load with the corresponding buckling mode. The term “eigenbuckling” comes from the method of formulation of the problem, so the eigenvalue equation for buckling problem take the form:

$$([K] + \lambda_i[S])\{\psi\}_i = \{0\}, \tag{4.34}$$

where [K] is a structural stiffness matrix, [S] is a stress stiffness matrix, λ_i is the i-th eigenvalue and ψ_i is the i-th eigenvector of displacement.

It should be added that the eigenvalue λ_i is a load multiplier that allows one to find the *i*-th buckling load. The *i*-th buckling load is calculated as the applied load multiplied by the *i*-th eigenvalue λ_i.

The stress stiffness matrix [S] is determined in the static analysis. It means that before the eigenbuckling analysis, the static analysis should be performed.

Equality (4.34) is satisfied if the eigenvector of displacement is equal to zero or if the determinant of the term in brackets is equal to zero. The $\{\psi\}_i = 0$—a trivial solution to (4.34)—is out of interest. It means that the structure does not change the "shape"—the structure remains in the initial state of equilibrium. Thus, the term in brackets in (4.34) gives the following solution:

$$|[K] + \lambda_i[S]| = 0. \tag{4.35}$$

Equation (4.35) represents the eigenvalue problem which allows for finding *n* values of the buckling load multiplier λ and the corresponding buckling mode shape. The number *n* depends on the number of DOFs assumed in the finite element model (*n* = DOF number).

It should be noted that in the FEM software as well as in the ANSYS software, the eigenvectors are normalized so that the largest component is 1.0. Thus, stresses may be interpreted only as a distribution of relative stresses.

4.4 Modal Analysis

The modal analysis is used to find natural frequencies and the corresponding modal modes. The solution to the problem is based on equation of motion (4.14) for an undamped system in which the left-hand side is equal to zero and which can be expressed in the form:

$$[\mathrm{M}] \cdot \{\ddot{\mathrm{u}}\} + [\mathrm{K}] \cdot \{\mathrm{u}\} = \{0\}. \tag{4.36}$$

The vector of the nodal displacement for free vibrations of a linear system is assumed as a harmonic of the following form:

$$\mathrm{u} = \{\phi\}_i \cos \omega_i t, \tag{4.37}$$

where $\{\phi\}_i$ is an eigenvector representing the mode shape of the *i*-th natural frequency, ω_i is the *i*-th natural circular frequency, and *t* denotes time.

Substituting the assumed vector of the nodal displacement (4.37) into equation of motion (4.36), the following relation is obtained:

$$\left(-\omega_i^2[M] + [K]\right)\{\phi\}_i = \{0\}. \tag{4.38}$$

Equality (4.38) is satisfied if the eigenvector representing the mode shape $\{\phi\}_i$ is equal to zero or if the determinant of the term in brackets is equal to zero. The

$\{\phi\}_i = 0$ is a trivial solution to (4.38), which is out of interest. Thus, the following solution is obtained:

$$\left|-\omega_i^2[M] + [K]\right| = 0. \tag{4.39}$$

Equation (4.39) represents the eigenvalue problem, which allows for finding n values of natural circular frequencies ω and the corresponding eigenvectors, which allow one to describe the modal mode. The number n equals to the number of DOFs assumed in the finite element model.

4.5 Element Type

The most suitable element for plated thin-walled structures is a shell element. An exemplary four-node shell element with six degrees of freedom at each node is presented in Fig. 4.4. In the case when the layered plate is modelled, a multilayer shell element can be used for which the material properties as well as orientations are defined for each layer.

Shell elements have only five independent degrees of freedom at each node (three perpendicular displacements and two rotations around the axis lying in the plane of the element). The sixth degree of freedom (a rotation around the axis normal to the plane of element) is not independent. Two different and possible shapes of displacements (a relation between rotations θ_z for each node in one element) are presented in Fig. 4.5. The relations between rotations are controlled for each element using the concept proposed by MacNeal and Harder [20].

The shape function for the presented element is assumed as the first order (linear) polynomial and has the following form:

$$N_e = \frac{1}{4}\left[u_I(1-s)(1-t) + u_J(1+s)(1-t) + u_K(1+s)(1+t) + u_L(1-s)(1+t)\right]. \tag{4.40}$$

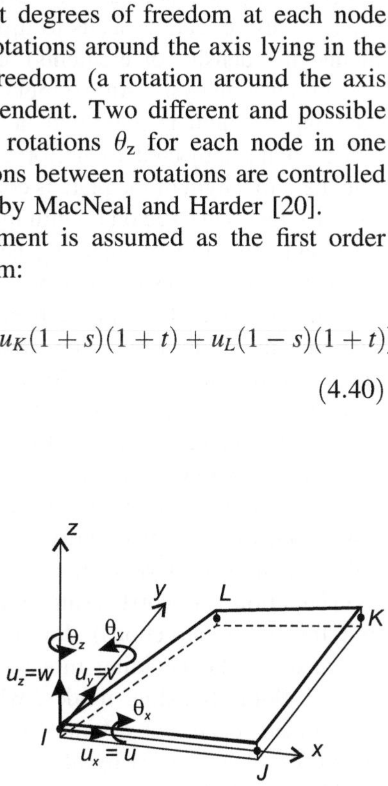

Fig. 4.4 Quadrilateral, four nodal shell element [1]

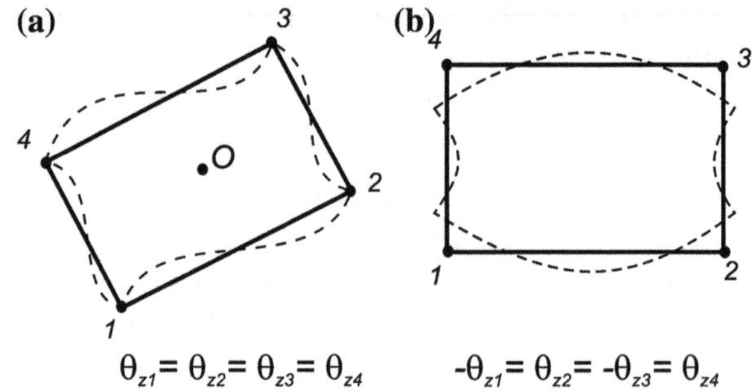

Fig. 4.5 Possible rotation around the axis normal to the plane of the element [1]. Spurious Mode (**a**), Hourglass Mode (**b**)

4.6 Discretization

As is very well known, the number of elements has a significant influence on the results of calculation. In all cases of the FEM calculations presented in this book, the number of elements has been assumed on the basis of experience [11, 12, 15–18]. The model discretized with a not enough number of elements is stiffer than the real structure and the results are not correct. The opposite case, i.e., too many elements used for discretization, leads to time-consuming calculations. The number of elements (the density of elements) should be chosen on the basis of the solution convergence analysis. An exemplary analysis of the solution convergence for the buckling problem is presented in Table 4.1. This analysis was performed for a square plate subjected to uniform compression. The plate thickness defined as thickness to length of the plate is equal to 1/100. The plate was simply supported on loaded edges and fixed on longitudinal edges. The use of four-node shell elements to represent the form of buckling needs the plate to be meshed into at least five elements along each sine halfwave. An increase in the number of elements increases the number of nodes and degrees of freedom for the model, which allows for mapping the highest buckling mode (the mode with more than one halfwave of sine). The 10 lowest buckling loads with the corresponding modes for different mesh densities (the number of elements in the longitudinal and transverse direction) have been determined and presented in Table 4.1.

It is well known that for the isotropic square plate fixed on the longitudinal edges and simply supported on the loaded edges, the mode corresponding to the lowest buckling has two halfwaves in the longitudinal direction ($m = 2$) and one halfwave in the transverse direction ($n = 1$). As presented in the exemplary results of calculation, the worst assumed mesh density (three elements in the longitudinal and transverse directions) did not give proper results—the lowest value was obtained for the buckling mode with one halfwave in both directions ($m = 1$, $n = 1$). Some buckling modes, which were found for a denser mesh, were not determined for the 3 by 3 elements mesh. An increase in the number of elements (from 3×3 to 5×5) by almost three times allows one to determine the same buckling mode (i.e., $m = 2$, $n = 1$) as is well known from the literature [25], but still not all buckling modes could be found (Table 4.1). Once again, an increased

Table 4.1 Influence of the mesh density on the buckling load [kN]

Mode[a] m	n	3 × 3	Change (%)[b]	5 × 5	Change (%)[b]	10 × 10	Change (%)[b]	20 × 20	Change (%)[b]	40 × 40
2	1	28.2	40	16.8	14	14.5	3	14.0	1	13.9
1	1	26.0	30	18.1	11	16.1	3	15.7	1	15.5
3	1	–		39.8	39	24.3	9	22.1	2	21.6
4	1	–		164.8	74	43.1	18	35.5	4	34.0
3	2	–		–		44.7	8	41.3	2	40.5
2	2	719.3	91	62.8	28	45.2	7	42.2	2	41.5
4	2	–		190.9	69	60.0	14	51.5	3	49.7
5	1	–		–		76.6	29	54.3	7	50.4
5	2	–		–		91.9	25	68.8	6	64.7
6	1	–		–		–		79.4	11	71.1

[a] *m, n*—number of halfwaves in longitudinal and transverse directions, correspondingly
[b] percentage change in the critical value relative to the result for the worse mesh

number of elements in the longitudinal and transverse directions (10 × 10) has resulted in further improvement of buckling loads value—the percentage change of the buckling load corresponding to all determined modes is in the range from 11 to 74 %. A 10 × 10 mesh allows one to determine nine from ten first modes which could be found using a denser mesh. A further increase in the number of elements (20 × 20) allows one to find all first ten modes and leads to a percentage change of the buckling load from 3 to 29 % in comparison with a less dense mesh. If the criterion that the improvement of the result by less than 5 % does not require a further increase of the mesh density is assumed, then it can be said that the appointment of the two lowest critical power density distributions of 20 × 20 is correct. Moreover, it can be noticed that the mesh improvement to receive 40 by 40 elements does not change first seven buckling loads by more than 5 % compared to the 20 × 20 mesh.

To be sure that the mesh density is correct not only for the static buckling analysis but also for the dynamic buckling analysis, the dynamic response for different amplitudes of the rectangular pulse load was checked. It is necessary because during and after the pulse load duration, the shape of deflection can change rapidly with an increase in the pulse amplitude.

For exemplary calculations, a rectangular pulse shape was chosen. The pulse duration is assumed to be equal to the period of natural vibrations for the analysed plate $T_p = T$. Two different mesh densities were considered—it was a 5 × 5 element mesh and a 20 × 20 element mesh in two orthogonal directions. Because the first modal mode has one halfwave of sine in the longitudinal direction ($m = 1$) and the first buckling mode has two halfwaves of sine, it was decided to conduct the analysis for both mode cases. For the case denoted as $m = 1$, the initial imperfection corresponded to the modal mode with the amplitude equal to 1/100 of the plate thickness, the pulse duration was equal to the period of natural vibrations for the mode with one halfwave of sine. The dynamic load factor DLF was calculated as the pulse amplitude divided by the buckling load for the mode $m = 1$ of the value corresponding to the assumed mesh density (Table 4.1). The similar assumptions were made for the case denoted as $m = 2$: the amplitude of initial imperfections equals 1/100 of the plate thickness and the initial imperfection shape corresponded to the first buckling mode ($m = 2$), the pulse duration equalled the period of natural vibrations for the mode $m = 2$ and the dynamic load factor DLF determined as the pulse amplitude divided by the buckling load for the mode $m = 2$.

The results are shown in Fig. 4.6 in the form of curves presenting the dependence of the maximum dimensionless deflection ξ as a function of the dynamic load factor DLF (Dynamic Load Factor—amplitude of the pulse load to the static buckling load). For the case denoted as $m = 2$, the curves ξ(DLF) presented in Fig. 4.6 for both mesh densities 5×5 and 20×20 are relatively close to each over. However, for the case of $m = 1$, the curves ξ(DLF) for both densities of the mesh taken into account overlap only for DLF ≤ 1.6 but for higher DLF values, the curves ξ(DLF) differ—see Fig. 4.6.

If the nodal displacement for $DLF = 3$ and for two mesh densities is compared (Fig. 4.7), it is clear that the model divided into a smaller number of elements (Fig. 4.7a) cannot present smoothly the shape of the deformed plate.

This example shows that the assumption of the finite element mesh density chosen in such a way that each square part of the column wall or the single plate is divided into 20×20 elements is correct and it should yield proper results.

4.7 Load and Boundary Conditions

Boundary conditions assumed on loaded edges depend on the analysed structure (a plate, a beam-column with an open and closed cross-section) and a type of load (uniform compression, eccentricity compression, pure bending). To obtain the results of the FEM analysis as close as possible to the experimental test or the numerical analysis made with another method, the assumed boundary conditions should be defined the same or similar in both compared tests/methods.

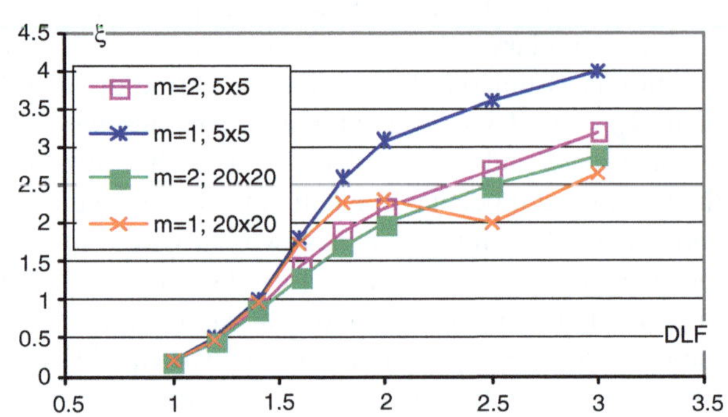

Fig. 4.6 Nondimensional deflection vs. the dynamic load factor for different elements sizes and buckling modes

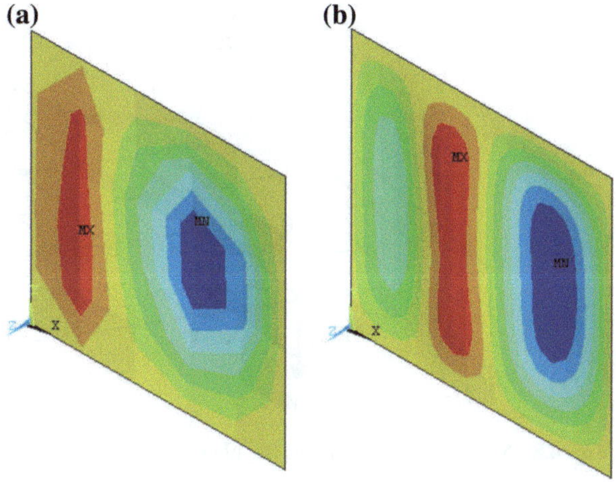

Fig. 4.7 Map of deflection for $DLF = 3$ and the element mesh equal to 5×5 (**a**) or 20×20 (**b**)

The boundary conditions in the FEM are determined defining nodal generalized displacements (displacements, rotations) and/or a coupled displacement for a group of nodes.

For all the analysed plates, the boundary conditions on longitudinal (unloaded) edges are set as follows (notation according to Fig. 4.8):

- simple supported: $v =$ constant and $w = 0$,
- clamped: $v =$ constant and $w = 0$ and $\theta_x = 0$,
- free edge—any constrains on the nodal displacement not set.

The boundary conditions set on loaded edges correspond to simply supported and, depending on a type of load, they are as follows:

- uniform compression: $u =$ constant and $w = 0$,
- pure in-plane bending or eccentricity compression: $\theta_z =$ constant and $w = 0$ and for one edge node lying on the neutral axis (middle node) $u = 0$.

To determine only the local buckling mode for the compressed column or girder, the boundary conditions assumed on loaded edges on the girder wall can be exactly the same as for the simply supported plate under uniform compression. In the case of pure bending or eccentricity compression of beam-columns or girders, the planes on which the loaded edges lie can rotate due to the assumed type of loading. Taking the above into account, the proper boundary condition should be considered. An exemplary way of finding the proper boundary condition for the girder with a closed cross-section is presented below [19].

The boundary conditions assumed in the analytical-numerical method in the prebuckling state in comparison to the buckling and postbuckling state are different. The above-mentioned differences were noticed and used in the FEM modelling of the compressed tube by Guarracino [8, 9]. It should be noted that also in the case of the girder with a closed cross-section subjected to pure bending, the boundary conditions in prebuckling and postbuckling states are different. That was the reason

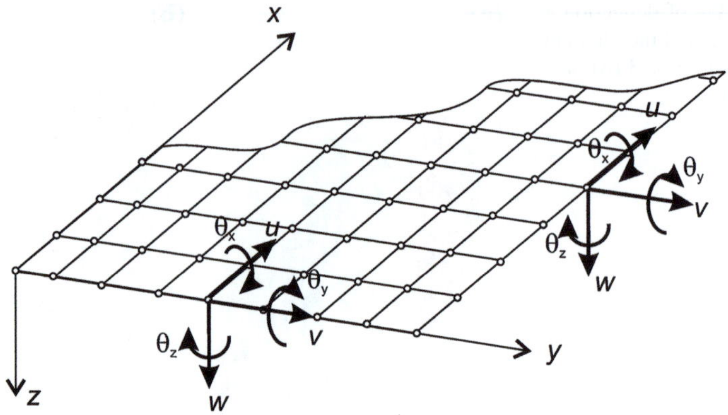

Fig. 4.8 Nodal displacement for the plate

to suggest and test different boundary condition models [18]. Three of them, denoted as **A**, **B** and **C,** respectively, are presented in Figs. 4.9, 4.10 and 4.11.

The proposed type **A** of boundary conditions (Fig. 4.9) ensures that the loaded edges remain straight—the nodal displacement for all nodes lying on wall edges in the direction normal to the wall was set to zero. In the next boundary condition model (type B—Fig. 4.10), the beam element was additionally applied as a stiffener on all loaded edges. The added stiffeners—the beam element—should be rigid for bending, compression and tension (the edges on which beam elements lie should stay straight after loading) and flexible for twisting (allow to rotate the plate or the wall around ending edges). To fulfil rigidity, the height of the beam is assumed to be 10 times greater than the thickness of the plate. To fulfil flexibility

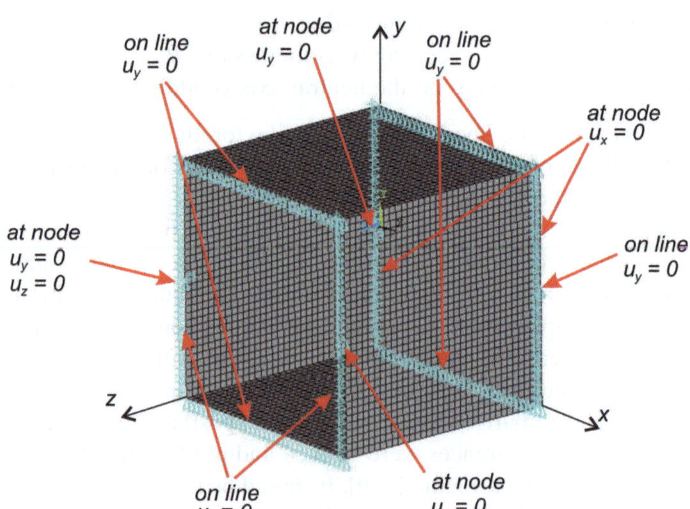

Fig. 4.9 Type **A** of boundary conditions—zero value of the nodal displacement

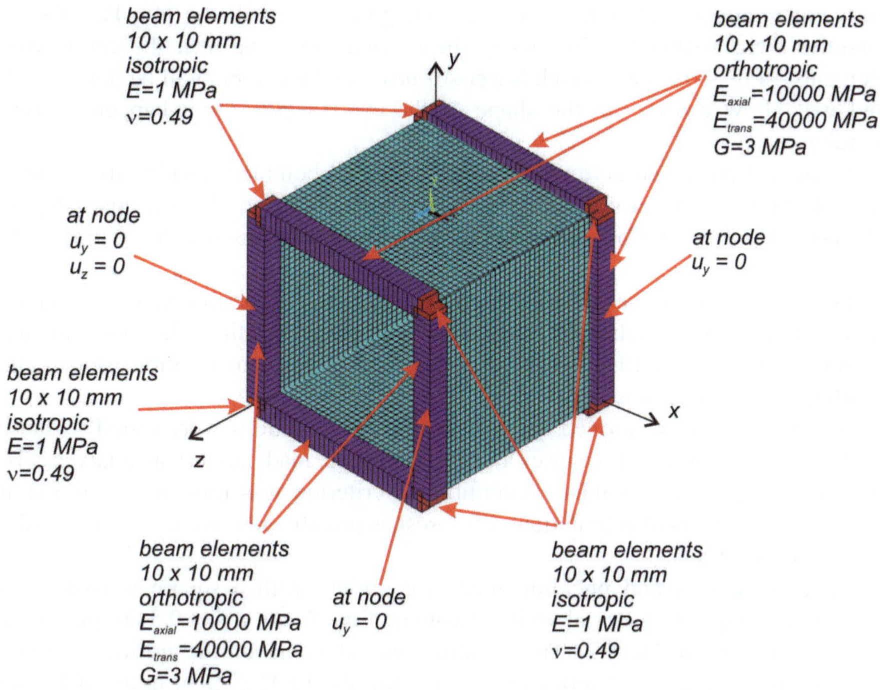

Fig. 4.10 Type **B** of boundary conditions—zero value of the nodal displacement and beam stiffeners

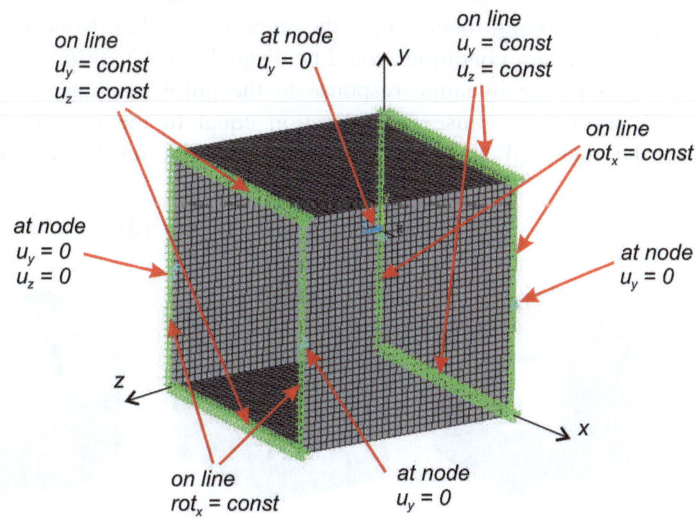

Fig. 4.11 Type **C** of boundary conditions—zero value of the nodal displacement and coupled boundary conditions

allowing rotations, the orthotropic material properties with the low Kirchhoff's modulus were assumed. Additionally, beam elements lying near the corner (red elements—Fig. 4.10) have much lower stiffness, so the corners will be not too stiff and will allow changes in the shape of the cross section as it happens during bending.

In the last presented here model of the assumed boundary conditions—type C (Fig. 4.11), not only zero value of the displacement but also of the coupled degree of freedom was used. For walls of girders or beam-columns which are uniformly compressed or tensioned, for all nodes lying on edges in longitudinal and normal to the wall directions, the constant displacement was set. For all nodes lying on the edges of the wall which were subjected to in-plane bending, the constant displacement in normal to the wall direction and a constant rotation around the bending neutral axis was set.

All three above-mentioned types of boundary conditions were tested in static buckling and dynamic responses during the pulse load numerical analysis. To choose "the proper" boundary conditions, a criterion was used which aimed at obtaining the results of calculations as closest as possible in both the methods—the FEM and the ANM.

Buckling modes and buckling loads for girders with a square cross-section made of a composite material with the volume fibre fraction $f = 0.5$ are presented in Fig. 4.12 and Table 4.2. The nondimensional value of the critical bending moment M_{cr}/M_{crANM} is determined by dividing the FEM critical moment by the critical moment obtained from the analytical-numerical method.

The assumed criterion and the results (Table 4.2) obtained from the stability analysis have shown that all the proposed boundary conditions can be considered to be correct—the differences between the results are less than 1 %. In all the analysed cases the buckling modes were the same (Fig. 4.12). None of the proposed models of boundary conditions could be disqualified. Therefore, for each of them, the analysis of the dynamic response to the pulse bending moment was carried out. A rectangular pulse with duration equal to the period of natural vibration of the analysed structure was assumed. The results are presented

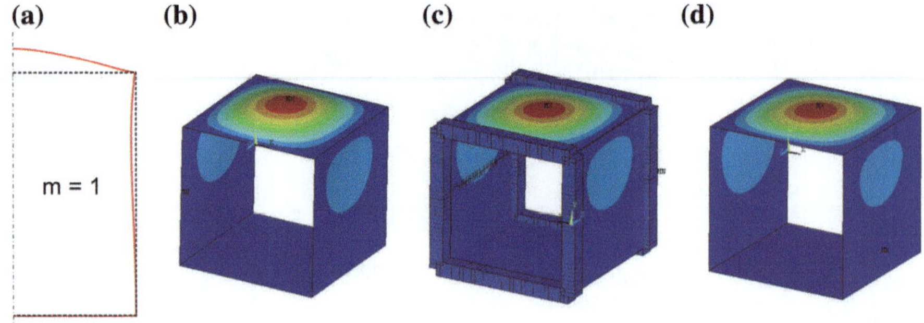

Fig. 4.12 Buckling mode obtained from the analytical-numerical method (**a**) and the finite element method with boundary conditions of types **A** (**b**), **B** (**c**) and **C** (**d**)

Table 4.2 Nondimensional buckling load comparison

	ANM	FEM		
		Type **A**	Type **B**	Type **C**
M_{cr}/M_{crANM}	1	0.991	1.001	0.991

(Fig. 4.13) in the form of the maximum deflection of the compressed wall of the girder as a function of the dynamic load factor $DLF = M_a/M_{cr}$, defined as a pulse amplitude of the applied loading divided by the static buckling load M_{cr}.

As can be seen from the curves presented in Fig. 4.13, considerable discrepancies in the results have been obtained with both methods. The differences increase with an increasing dynamic load factor value, which for a higher load lead to larger deformations. The applied boundary conditions are much more significant than for small deformations. Differences in the obtained results mean that the boundary conditions were not properly chosen, which can be confirmed after a closer examination of possible displacements of loaded girder edges. The end section of the bent girder rotates around the neutral axis, and therefore the edges of the upper and lower walls cannot only move along the beam axis but also in the direction normal to the surface of the flanges. Such movements are allowed in the prebuckling state in the analytical-numerical method, however are not possible in the finite element model (Fig. 4.9). Accordingly, the boundary conditions **A** are disqualified and a further analysis was performed for comparison of the two other ways of support (case **B**—Fig. 4.10 and case **C**—Fig. 4.11). As is well visible in Figs. 4.10 and 4.11, the boundary conditions **B** and **C** allow for the necessary movement of ends of the girder subjected to the bending moment.

Analysing the results presented in Fig. 4.13, one can conclude that the best boundary conditions allowing to map and verify the model assumed in the analytical-numerical method are conditions **C** (Fig. 4.11).

Comparing the results of the ANM and the FEM with **A** boundary conditions, it can be seen that the deflection differences increase with an increasing DLF, which means that the boundary conditions make the model stiffer, especially for large deformations. Also, the curve obtained for the **B**-type boundary conditions is far

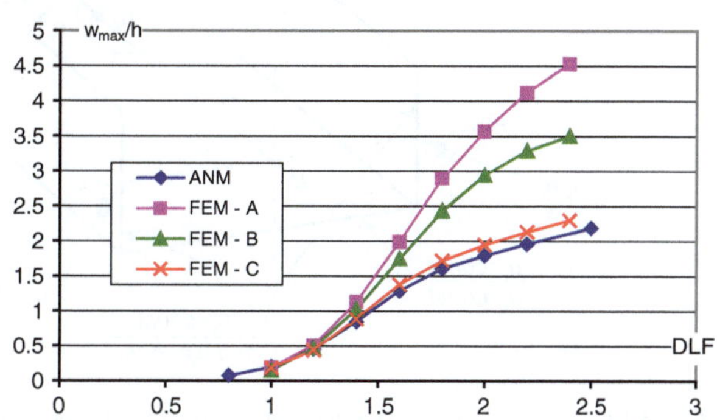

Fig. 4.13 Dimensionless deflection versus DLF for different types of the boundary conditions assumed in the FEM and the ANM

from the curve of the analytical-numerical method. Beam elements in the model **B** do not have an infinite stiffness and, therefore, cannot ensure that the ending cross-section remains flat and the loaded edges of the walls remain straight. The **C**-type boundary conditions allow for free rotation of the final section (which remains flat) around the neutral axis. Thus, for a further finite element analysis (numerical experiment) of girders subjected to bending, the **C**-type boundary conditions are adopted (Fig. 4.11).

Therefore, for all columns with closed cross-sections subject to compression, in which the global mode was taken into consideration, the boundary conditions analogical to the **C**-type were assumed. A similar analysis as the one described above may be carried out for the beam-column with an open cross-section for which the boundary conditions in loaded ends were assumed according to the description presented in Fig. 4.14.

Figure 4.14 shows one example of a column with an open cross-section. For other types of open cross-sections, the boundary conditions were assumed by analogy. It was assumed that bending (the flexural buckling mode) occurred around the axis for which the second moment of area is the smallest, thus the FEM model was prepared in such a way that nodes in which the displacement in the y direction was set to zero (Fig. 4.14) were on the neutral axis of ending sections. Straightness of the loaded edges of the considered beam-column is provided by requiring equal displacements of all nodes lying on the edge of the beam-column in the direction normal to its walls. To ensure that deformations are compatible with the deflection in bending (the global flexural buckling mode), the edges normal to the neutral axis remained straight in the plane containing the wall of the column. In addition, for all nodes lying on those edges, the constant rotation around the axis parallel to the axis of the neutral section was presupposed.

When the compressed column was analysed, a compression force or a uniform stress distribution with additional conditions for the uniform edge displacement was assumed. In other cases of load (pure bending or bending with compression),

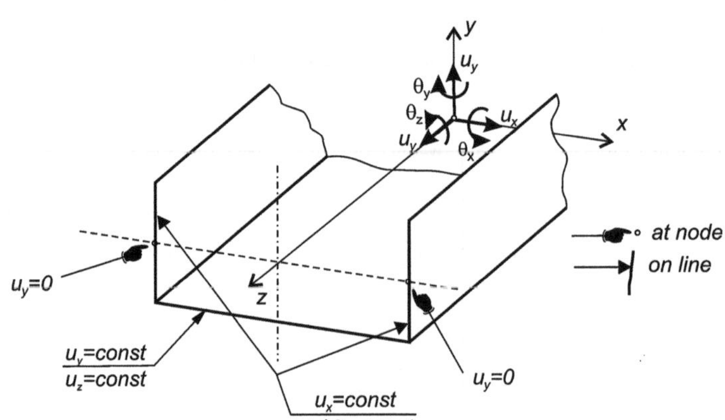

Fig. 4.14 Assumed boundary conditions—an example for a channel cross-section

the stress distribution corresponding to a given load was assumed. The stress distribution was modelled in the form of the pressure distribution acting along the loaded edges of the structure. An exemplary model of load corresponding to pure bending is presented in Fig. 4.15.

It should be noted that the assumed load is nonconservative—pressure is always perpendicular to the area to which it is applied. Let us check what an influence of the assumed model, i.e., the load and the boundary condition, on the results of calculations is. As an example, the channel-shape cross-section profile subjected to pure bending was computed. Four different models denoted as BC-1 to BC-4 were considered, and they are as follows:

- BC-1 part (with a length *l*) of the beam-column between two neighbouring diaphragms, the load is a pressure modelling the bending stress distribution; the boundary condition (displacement set to zero) assumed in neutral axes; in loaded edges of the beam-column, a constant value of displacement in the normal direction to its wall is assumed—Fig. 4.16a;
- BC-2 BC-1 model with added constant displacement in the longitudinal direction of the beam-column loaded edges of the web—Fig. 4.16b;
- BC-3 considered part of the beam of a length *l* modelled together with a handle subjected to four-point bending (Fig. 4.17);
- BC-4 model three times longer than the considered part of the beam with a diaphragm (Fig. 4.18) and a handle, the whole subjected to four-point bending.

For all the above-mentioned models of static buckling, a modal and nonlinear static analysis was performed. The results of these calculations are presented in Tables 4.3 and 4.4 and in Figs. 4.19 and 4.20.

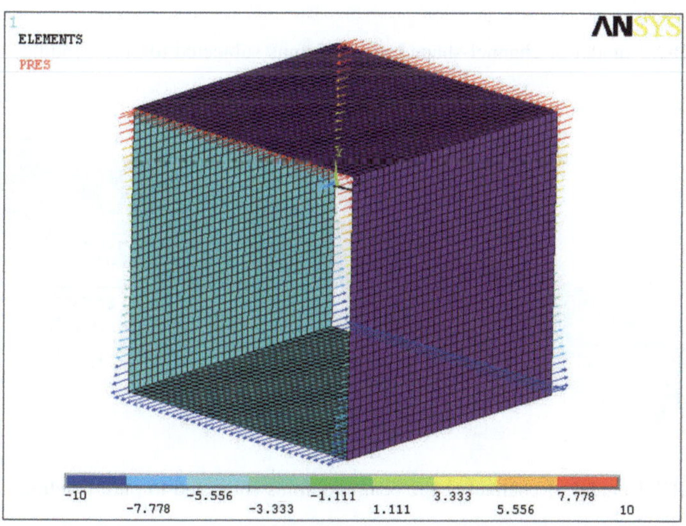

Fig. 4.15 Exemplary load model for a segment of the girder subjected to pure bending

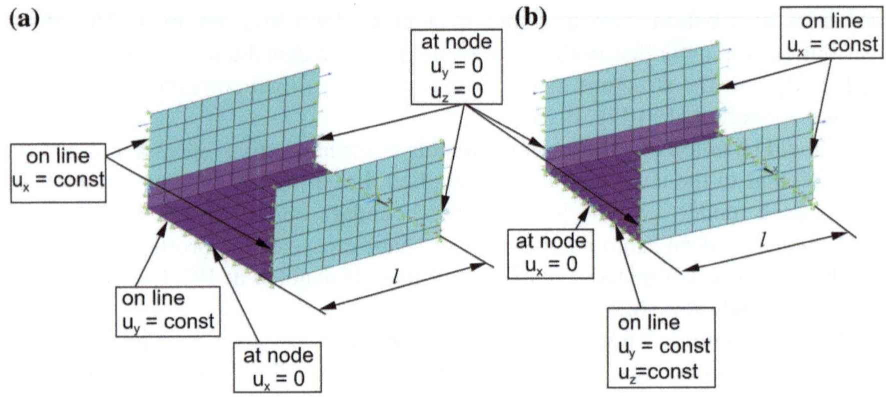

Fig. 4.16 BC-1 (**a**) and BC-2 (**b**) models of channel-shape beam-columns subjected to pure bending

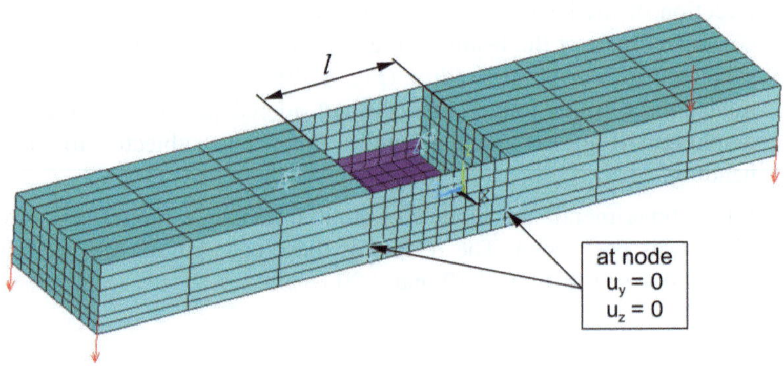

Fig. 4.17 BC-3 model of channel-shape beam-columns subjected to pure bending

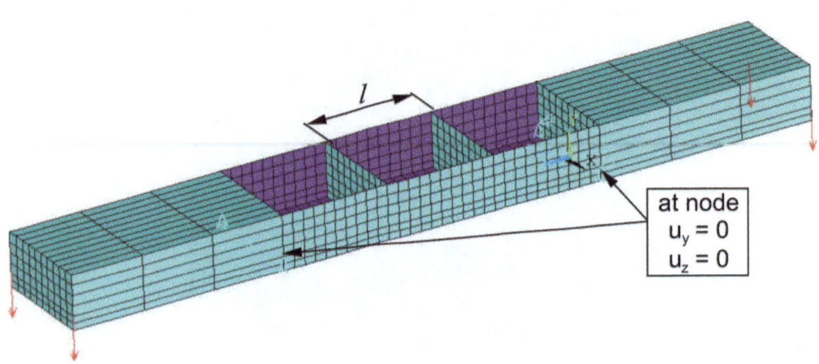

Fig. 4.18 BC-4 model of channel-shape beam-columns subjected to pure bending

Examining the results from the eigenvector analysis (the modal analysis—Table 4.3 and the buckling analysis—Table 4.4), it is well visible that all four considered models of boundary conditions and loads do not play a significant role in the results of calculations—the highest differences are less than 5 %. It means that the simplest models (BC-1 and BC-2) are accurate enough for the static linear analysis. It should be noted that the lowest natural frequencies and the corresponding modal modes for the model denoted as BC-4 are different than for BC-1 and BC-2. The modal mode differences are shown in Fig. 4.19. As can be seen in Fig. 4.19, the first modal mode obtained using the BC-4 model has a global character (Fig 4.19b). Due to these differences, in Table 4.3 only the corresponding modal modes are compared (for example, the first from BC-1 presented in Fig. 4.19a with the third one from BC-4 presented in Fig. 4.19c).

Comparing the results of the nonlinear buckling analysis, it is clear that the considered models of boundary conditions and loads have a significant impact on the structure work in the postbuckling state (Figs. 4.20 and 4.21). Some similarities can be seen when the postbuckling behaviour for the models denoted as BC-2 and BC-4 is compared. The first model denoted as BC-1 is too 'weak'. The lack of the straightness assumption for the loaded web edge leads to a reduction in its stiffness. The model designated as BC-3 is too 'stiff' due to the close box neighbourhood to the considered beam-column part and a much stiffer handle. To be sure which model of load and boundary conditions is correct, experimental tests should be performed.

The next considered case is a girder subjected to torsion. To introduce the load causing twist, one end of the girder was restrained (for all nodes, three perpendicular displacements were set to zero) and two pairs of forces were applied to the second end (Fig. 4.22). Such an assumption leads to a deformation on the not restrained end of the girder and to a not natural stress concentration in the places where the forces are applied (Fig. 4.23).

Table 4.3 Natural frequencies for different models of load and boundary conditions

Length l (mm)	Natural frequencies f (Hz)				Differences[a] (%)
	BC-1	BC-2	BC-3	BC-4	
40	1,117	1,117	1,118	1,135	1
50	836	836	836	850	2
60	678	678	678	690	2
70	579.4	579.4	580	573	1
80	513	513	513	516	1
90	466	466	466	469	1
100	431.3	431.3	431	434	1
150	343.8	343.8	344	345	0
200	310	310	311	310	0
250	293.5	293.5	295	287	3

[a] Differences calculated between the minimal and maximal value divided by the maximal value

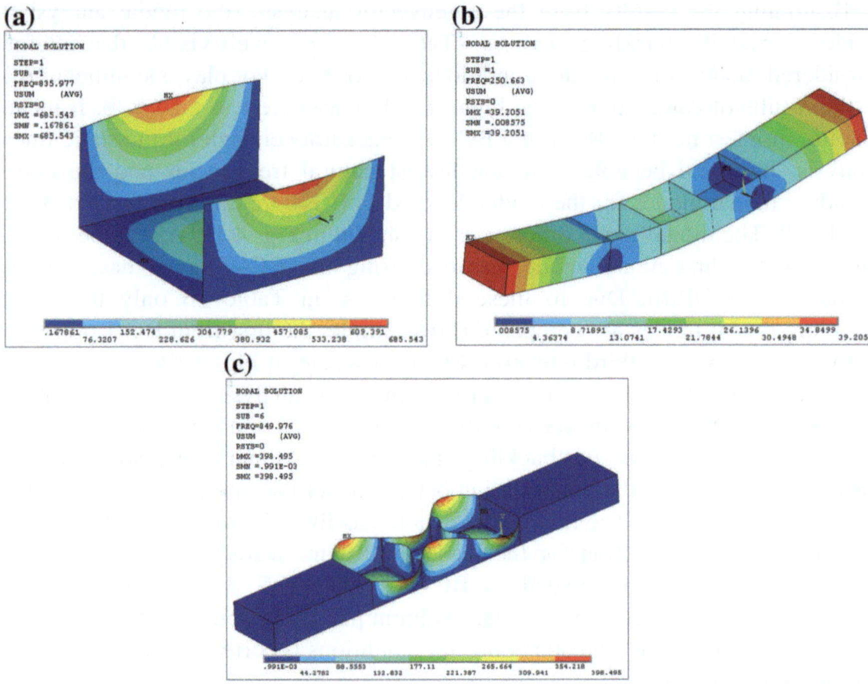

Fig. 4.19 First modal mode for BC-2 (**a**) and BC-4 (**b**) models of the beam-column with the length $l = 50$ mm

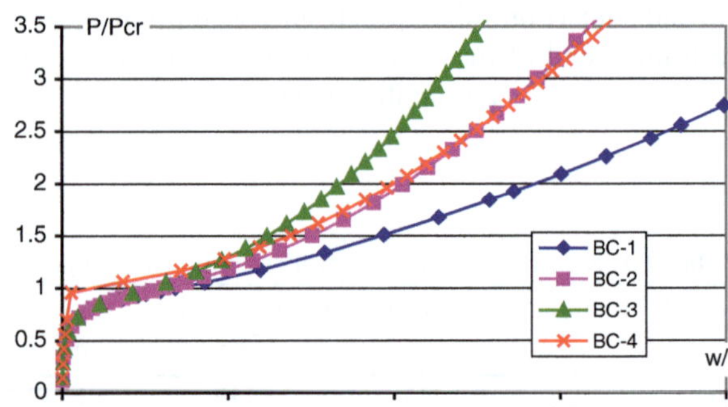

Fig. 4.20 Postbuckling equilibrium path for the channel-shape beam-column of the length $l = 50$ mm

To prevent a stress concentration at loaded points and a deformation of the shape of the one end of the girder, the front panel 10 times thicker than the girder wall thickness (Fig. 4.24) was introduced.

The proposed model can be employed in the analysis of girders subjected to torsion because there are no unexpected deformations—both ends of the girder after loading remain the same square shape (Fig. 4.25).

The most difficult arrangement of the boundary condition was for the girders subjected to a combined load—a simultaneously applied bending moment and a

Table 4.4 Critical bending moment for different models of boundary conditions

Length l (mm)	Buckling mode (m)	Critical bending moment M_{cr} (Nm)				Differences[a] (%)
		BC-1	BC-2	BC-3	BC-4	
40	1	52	52	51	53	4
50	1	55	55	55	57	3
60	1	63	63	62	65	4
	2	57	57	58	58	3
70	1	73	74	73	76	3
	2	53	53	53	54	2
80	1	86	86	87	88	2
	2	52	52	53	53	3
90	1	101	101	102	104	2
	2	53	53	54	54	2
100	1	118	118	120	121	2
	3	54	54	55	55	2
150	1	226	226	233	233	3
	4	52	52	53	54	3
200	1	359	360	372	374	4
	5	52	52	53	53	2
250	1	493	495	516	514	4
	6	52	52	53	55	5

[a] Differences calculated between the minimal and maximal value divided by the maximal value

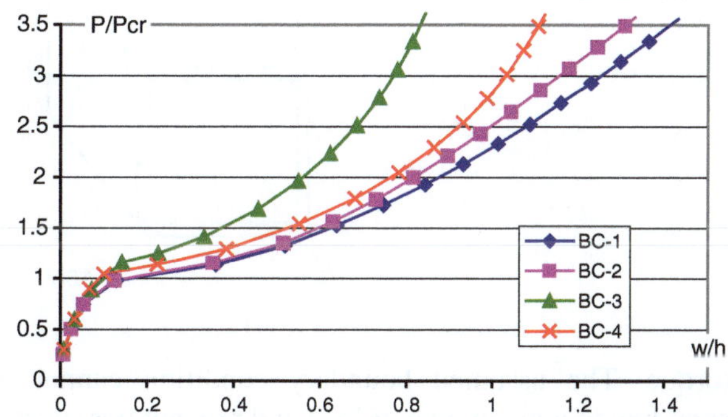

Fig. 4.21 Postbuckling equilibrium path for the channel-shape beam-column of the length $l = 200$ mm

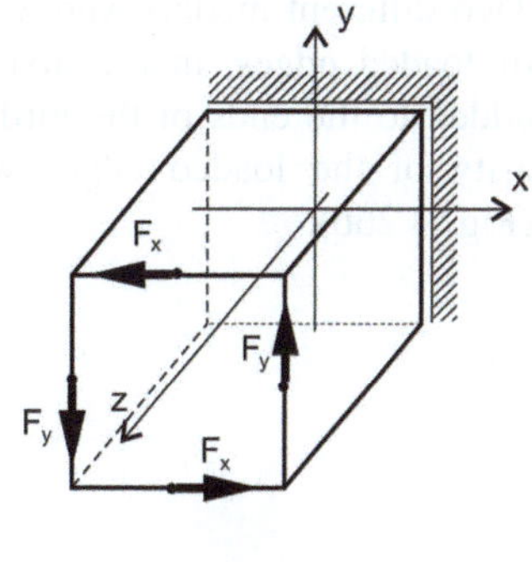

Fig. 4.22 Model of the boundary conditions and the load for the analysed girder

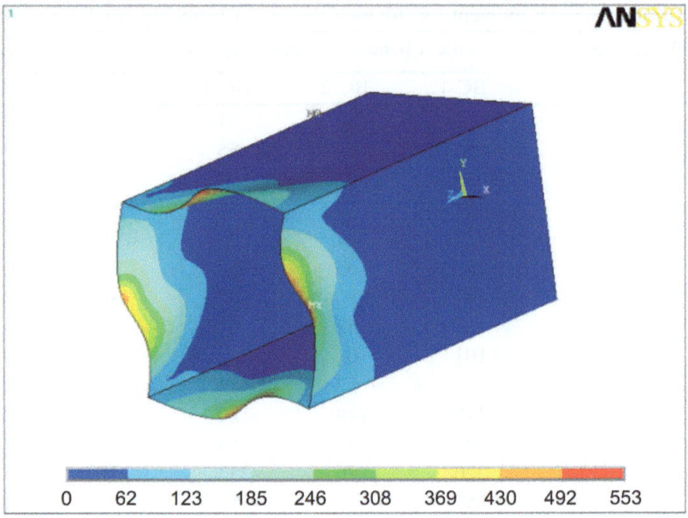

Fig. 4.23 Deformation and the equivalent stress distribution for the first model

Fig. 4.24 Models with a solution preventing deformation and stress concentrations

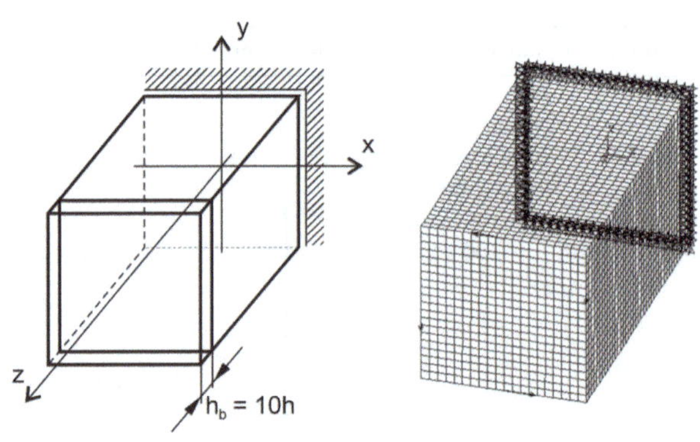

torque. The assumed boundary condition cannot block displacements corresponding to a given type of load and prevent unexpected stress concentrations (i.e., the stress concentration in the corners on the plane where the loads are applied). Two different models were considered (Fig. 4.26). In order to assure the linearity of loaded edges, in the first model two plates of relatively high stiffness were added to the ends of the girder (Fig. 4.26a). In the second model, the same linearity of the loaded edges was obtained by an application of beam elements (Fig. 4.26b).

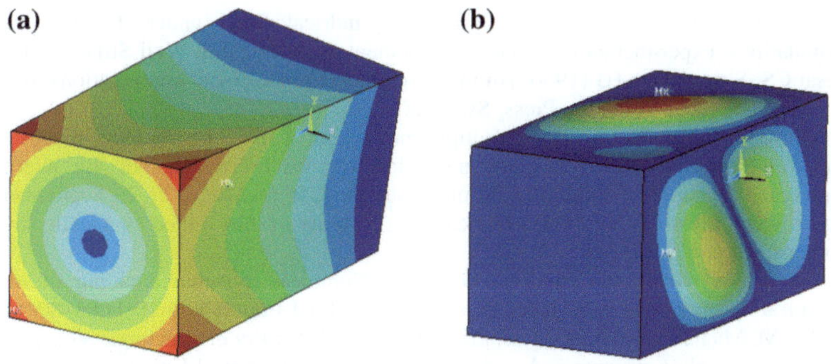

Fig. 4.25 Results of static (**a**) and buckling (**b**) analyses for the model with a front panel

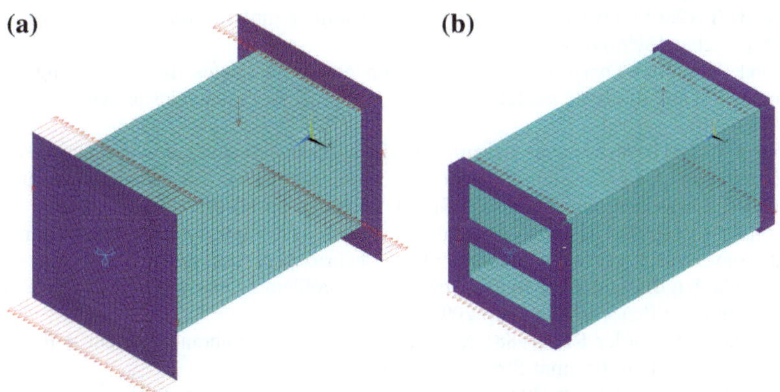

Fig. 4.26 FEM models: **a** plate model (Model 1), **b** beam model (Model 2)

References

1. ANSYS 11.1 html online documentation, SAS IP, Inc, 2007
2. Bathe KJ (1996) Finite element procedures. Prentice Hall, Englewood Cliffs
3. Chroscielewski J, Makowski J, Pietraszkiewicz W (2004) Statyka i dynamika powlok wieloplatowych: nieliniowa teoria i metoda elementow skonczonych. IPPT PAN, Warsaw
4. Courant R (1943) Variational methods for the solution of problems of equilibrium and vibrations. Bull Am Math Soc 49:1–23
5. Crisfield MA (1997) Non-linear finite element analysis of solids and structures. John Wiley & Sons, Inc, New York
6. Forde WRB, Stiemer SF (1987) Improved arc length orthogonality methods for nonlinear finite element analysis. Comput Struct 27(5):625–630
7. Fortuna Z, Macukow B, Wąsowski J (2006) Metody Numeryczne. WNT, Warsaw
8. Guarracino F (2001) The elastic-plastic behaviour of cylindrical tubes under bending: finite elements analyses, experimental data and theoretical consideration. In: Zaras J, Kowal-Michalska K, Rhodes J (eds) Proceedings of international workshop on the application of finite element method to the analysis of thin-walled structures, addendum to proceedings of third international conference of thin-walled structures, Cracow, pp 5–14

9. Guarracino F (2003) On the analysis of cylindrical tubes under flexure: theoretical formulations, experimental data and finite element analyses. Thin Wall Struct 41:127–147
10. Desai CS, Siriwardane HJ (1980) An introduction to finite element computations. In: Hinton E, Owen DR (eds) Pineridge Press, Swansea
11. Kolakowski Z, Kubiak T (2004) Multiple interaction of dynamic buckling modes in thin-walled members subjected in-plane pulse loading. In: Proceedings of 4th international conference on coupled instabilities in metal structure, Rome, Italy, 27–29 September 2004
12. Kolakowski Z, Kubiak T (2005) Load-carrying capacity of thin-walled composite structures. Compos Struct 67:417–426
13. Kounadis AN, Gantes C, Simitses G (1997) Nonlinear dynamic buckling of multi-dof structural dissipative system under impact loading. Int J Impact Eng 19(1):63–80
14. Krolak M, Mania RJ (eds) (2011) Static, dynamic and stability of structures. Vol. 1: Stability of thin-walled plate structures. A series of monographs. Technical University of Lodz Press, Lodz
15. Kubiak T (2005) Dynamic buckling of thin-walled composite plates with varying widthwise material properties. Int J Solids Struct 45:5555–5567
16. Kubiak T (2005) Dynamic buckling of thin-walled girders with channel cross-section. VDI-Berichte 1899:69–78
17. Kubiak T (2005) Dynamic buckling of thin composite plates. In: Proceedings IUTAM symposium on multiscale modelling and fracture processes in composite materials, Springer, pp 123–130
18. Kubiak T (2005) Wyboczenie dynamiczne cienkosciennych dzwigarow kompozytowych. In: Proceedings of IX-th scientific-technical conference FEM software in computer aided analysis, design and manufacturing, Gizycko, Poland, pp 89–96
19. Kubiak T (2007) Metoda elementow skonczonych jako eksperyment numeryczny statecznosci dzwigarow cienkosciennych obciazonych statycznie i dynamicznie. In: Niezgoda T (ed) Numerical analysis of selected problem in mechanics. Military University of Technology Press, Warsaw, pp 209–228
20. MacNeal RH, Harder RL (1988) A refined four-noded membrane element with rotational degrees of freedom. Comput Struct 28(1):75–84
21. Reddy JN (2004) An introduction to nonlinear finite element analysis. Oxford University Press, New York
22. Riks E (1979) An incremental approach to the solution of snapping and buckling problems. Int J Solids Struct 15:529–551
23. Szmelter J, Dacko M, Dobrocinski S, Wieczorek M (1973) Komputerowe programy metody elementow skonczonych. Arkady, Warszawa
24. Szmelter J, Dacko M, Dobrociński S, Wieczorek M (1979) Metoda elementow skonczonych w statyce konstrukcji. Arkady, Warszawa
25. Timoshenko SP, Gere JM (1961) Theory of elastic stability. McGraw-Hill, New York
26. Turner MJ, Clough RW, Martin HC, Topp LJ (1956) Stiffness and deflection analysis of complex structures. J Aero Sci 23(9):805–823
27. Washizu K (1974) Variational method in elasticity and plasticity. Pergamon Press, Oxford
28. Zienkiewicz OC (1971) The finite element method in engineering science. McGraw-Hill, London
29. Zienkiewicz OC, Cheung YK (1967) The finite element method in structural and continuum mechanics: numerical solution of problems in structural and continuum mechanics. McGraw-Hill, London
30. Zienkiewicz OC, Taylor RL (2000) The finite element method. Butterworth-Heinemann, Oxford

Chapter 5
Dynamic Buckling Criteria

Dynamic stability, or better, dynamic buckling (sometimes referred to as a dynamic response) means a loss of stability of the structure subjected to pulse load. Especially, it can act along the axis of the column or in the plane of the plate. It should be mentioned that for the ideal uniformy compressed structures (without any geometrical imperfection) the critical buckling amplitude of pulse loading leads to infinity. Therefore, the dynamic buckling can be analysed only for structures with initial geometrical imperfections. In such a case, the critical value of dynamic load does not have a bifurcation character, thus it has to be determined on the basis of the assumed criterion. Some the most popular and some new criterion allowing to determine dynamic buckling is presented in this chapter.

There are authors (e.g. [22]) who claim that for the thin-walled structures with flat walls (structures with stable postbuckling equilibrium path) dynamic buckling does not occur—the dynamic response can be analyse only.

Raftoyiannis and Kounadis [20] divided the criteria into the geometric and energetic ones. Geometric criteria are those in which a loss of dynamic stability refers usually to deflection or shortening, whereas the energy criterion is the one in which the critical value is determined by the potential and/or kinetic energy of the system.

Another division of criteria distinguishes the criteria for structures with a stable postbuckling equilibrium path (plate structures) and the criteria for structures with an unstable postbuckling equilibrium path or having a limit point (shells, rods). For structures with an unstable postbuckling equilibrium path, it is possible to derive mathematically the relation allowing for finding the critical value, describing the dynamic buckling load [11]. However, for plate structures the mathematical derivation is not possible and the criteria have been formulated from observations of behaviour of such structures [2] or on the basis of experiments [1, 6, 8]. Therefore, from the moment when the dynamic buckling problem appears in the literature, a number of dynamic stability criteria has been developed.

In the literature, in addition to displacement and energy criteria, failure criteria can be found. Such a criterion has been formulated by Petry and Fahlbusch [19]. In failure criteria for the dynamic buckling stability, proper hypotheses to determine

T. Kubiak, *Static and Dynamic Buckling of Thin-Walled Plate Structures*,
DOI: 10.1007/978-3-319-00654-3_5,

the equivalent stress state are needed. The equivalent stress state can be compared with the following material properties: yield limit or ultimate stress for tension or compression. It should be noted that in the majority of works (also here), the assumed material properties were obtained from static tests.

5.1 Volmir Criterion

Volmir in his work [28] presented the behaviour of a simply supported rectangular plate subjected to different pulse loads. He analysed the pulse of infinite duration, linearly increasing load and pulses of finite duration. The pulses taken into consideration had rectangular and exponentially decreasing shapes. Volmir solved the dynamic buckling problem using the Bubnov-Galerkin method for determining the buckling and postbuckling state for statically loaded structures. Then, after setting the deflections and load as a function of time, the equations of motion were obtained. The equations of motion are solved using the Runge-Kutta method. During his analysis, Volmir took into consideration not only tensile load pulses but also the shear type. Probably due to computational difficulties, Volmir proposed a very simple but time-consuming method for determining the "critical" dynamic load. This method consists in assuming the buckling mode and dynamic response analysis. This assumed buckling mode (a number of sine halfwaves describing out-of-plane deflection) was taken as the critical one for uncoupled dynamic buckling if the dynamic response of the plate subjected to a given amplitude of pulse resulted in an increase of deflection in the shortest time. Volmir considered the buckling problem which can be described by a theoretical system with one degree of freedom. For the critical mode, "a factor of dynamism K_D" (further referred to as the Dynamic Load Factor—DLF) was determined. The K_D factor was defined as a ratio of the pulse amplitude of the critical load to the static critical load. On the basis of his study, Volmir suggested a very simple criterion for the dynamic stability loss, assuming that a loss of stability of the plate subjected to pulse load occurs when the maximum deflection of the plate is equal to the assumed constant value. Usually the critical deflection value was assumed to be equal to the thickness of the plate or half of its thickness.

5.2 Budiansky–Hutchinson Criterion

One of the first displacement criteria was formulated by Budiansky and Hutchinson [4, 11]. This criterion involves structures with geometrical imperfections and an unstable postbuckling equilibrium path or a limit point. The authors of that criterion analysed cylindrical shells and axially loaded rods. A similar criterion was formulated for cylindrical shells loaded transversely by Budiansky and Roth [5]. They considered pulse load of finite or infinite duration and derived the relationship which

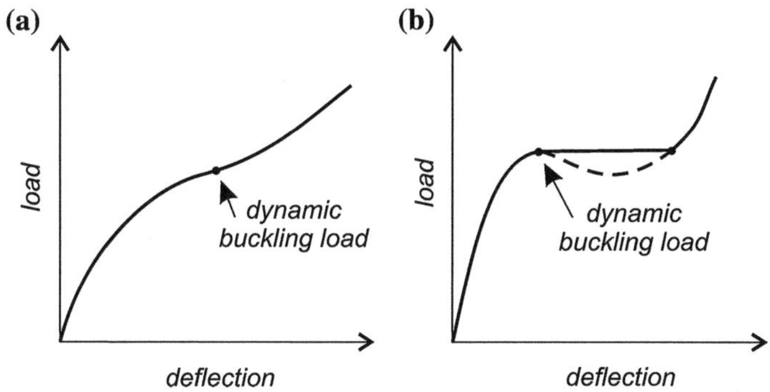

Fig. 5.1 Equilibrium paths for pulse loaded structures with large (**a**) and small (**b**) initial geometrical imperfections

allows for determining the critical load. It has been noted that the critical load corresponds to the inflection point on the curve presenting the load as a function of deflection (Fig. 5.1). Budiansky and Hutchinson defined their criterion in the following form:

A loss of stability of structures subjected to pulse loading occurs when there is an unlimited increase of deflection for small increments of load.

Many authors have adopted the above criterion for plated structures.

On the basis of the time history of the structure deflection, for a given pulse shape and its duration, a graph of the maximum deflection amplitude as a function of the load is built (Fig. 5.2). The Budiansky-Hutchinson criterion adopted for plated structures states:

A dynamic stability loss occurs when the maximum plate deflection grows rapidly with a small variation of the load amplitude.

5.3 Ari–Gur and Simonetta's Criterion

Ari-Gur and Simonetta [2] conducted a series of experiments and theoretical analyses of thin plates clamped on all edges and subjected to pulse load with a halfwave of sine shape (with finite duration). They noted that for a perfectly flat plate, the pulse load intensity L (the force F or the shortening U), which would result in a loss of stability, was infinitely large. The Ari-Gur and Simonetta's critical load value was set as depending on the following parameters:

- deflection—measured in the middle of the length and the width of the plate,
- load intensity L_m defined as the force pulse amplitude F_m or the shortening U_m.

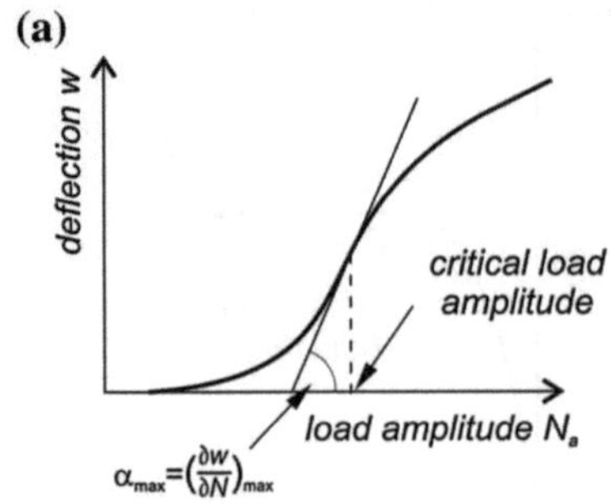

Fig. 5.2 Typical curves representing deflections as a function of load for the local (**a**) or global (**b**) buckling mode

On the basis of the analytical and numerical study, four dynamic buckling criteria are proposed in [2]. The first one (Fig. 5.4a) is based on the observation of the deflection w_m (Fig. 5.3) and the intensity of the pulse load L_m. It was formulated as follows:

Dynamic buckling occurs when a small increase in the intensity of the pulse load L_m causes a significant increase in the value of the deflection w_m.

The above criterion is analogous to the criterion defined by Budiansky and Hutchinson (Sect. 5.2). Ari-Gur and Simonetta have noticed that the criterion can be used for loads in the form of a force or a displacement and for different pulse load durations. However, for a very short period of the pulse duration, a very high amplitude value in deflections is required (a large amplitude in comparison to the static critical load) to obtain a rapid increase, which—as is known—can cause a change in the mode of deflections during the pulse load. These observations have led to the formulation of the second criterion which is based on the analysis of the maximum value of the pulse load L_m and the deflection value w_m (Fig. 5.4b). This criterion is as follows:

Dynamic buckling occurs when a small increase in the amplitude of the pulse load L_m causes a decrease in the value of the deflection w_m.

The next two criteria are failure criteria which are based on a response of the loaded edge of the considered plate, namely shortening (for a force as a pulse) or

Fig. 5.3 Analysed pulse load and measured parameters

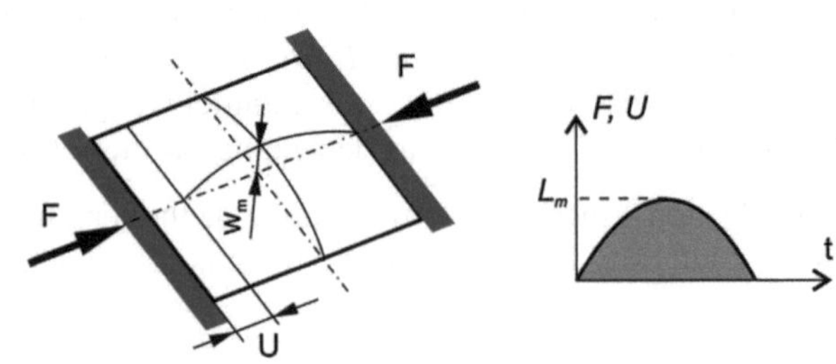

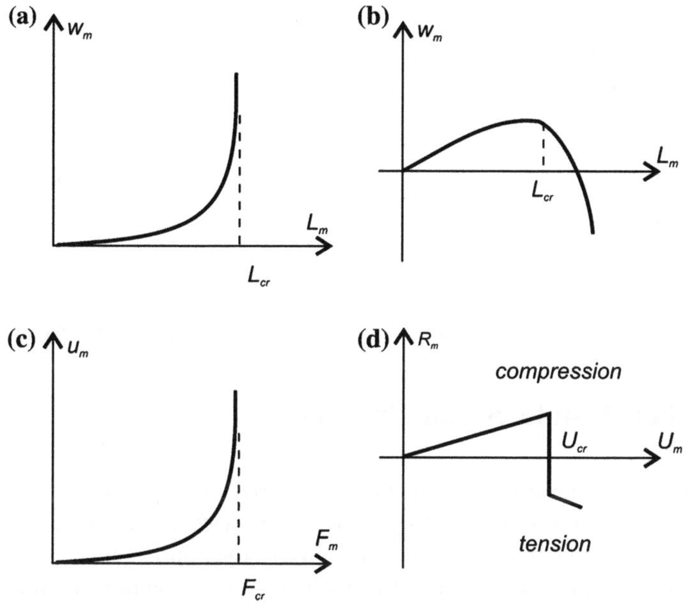

Fig. 5.4 Graphs presenting Ari-Gur and Simonetta's dynamic buckling criteria

reaction (for a displacement as a pulse load) on loaded edges. In the case of the shortening analysis, a similarity to the increasing deflection analysis (the first Ari-Gur and Simonetta's criterion) can be noticed. The third criterion (Fig. 5.4c) was formulated as follows:

> *Dynamic buckling occurs when a small increase in the amplitude of the force pulse F_m causes a sudden increase in the shortening u_m of loaded edges of the plate.*

A significant increase in deflection of the plate will reduce its stiffness, so the load which causes a significant increase in the shortening is defined as the critical load (Fig. 5.4c).

The last criterion applies to the case when the load is defined not by a force but by an impulse of displacement (shortening of loaded edges) and defines the critical U_m displacement pulse intensity. The fourth criterion (Fig. 5.4d) given by Ari-Gur and Simonetta is:

> *Dynamic buckling occurs when a small increase in the pulse displacement intensity U_m of the loaded edge causes a change in the sign of the value of the reaction R_m at the edge of the plate.*

As is well known for the deflected plate, the reaction distribution on the loaded edge has a sinusoidal shape (Fig. 5.5), thus in the case of large deflections, the tension appears in the middle of the loaded edge which maintains the straightness of the edge under load. The value of the resultant tensile force can be greater than the compressive forces that occur outside the central part of the edge under load.

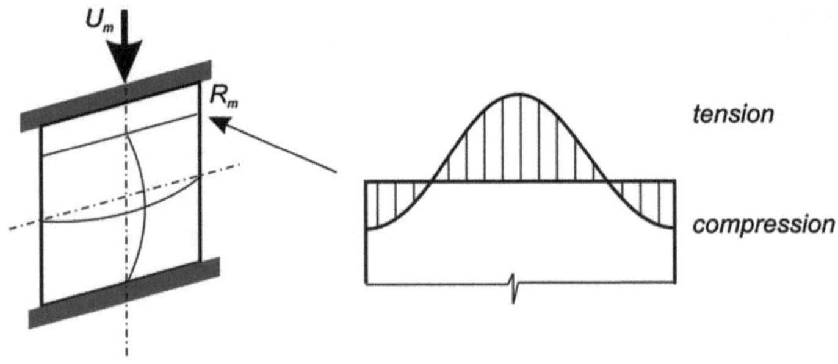

Fig. 5.5 Force distribution along loaded edges of the plate with initial imperfections

5.4 Kleiber–Kotula–Saran Criterion

Kleiber, Kotula and Saran in their work [13] presented the problem of dynamic stability for rod systems. The problem was solved with the finite element method. The paper quoted the criterion of stability in the Lyapunov sense, which requires disturbing the initial conditions and the analysis of the system behaviour. The aim of the mentioned work [13] was to formulate such a criterion that would allow the dynamic stability analysis based only on solving basic equations. It is well known that FEM solutions for static loaded structures are unstable when the tangent stiffness matrix is singular. Kleiber, Kotula and Saran began to analyse the tangent stiffness matrix for a system of rods subjected to pulse load with infinite duration (Heaviside pulse loading). First, they defined the critical pulse duration t_{cr}. For the time t less than the critical one, structural deflections are very small and oscillate around the position of equilibrium. When the time is longer than the critical t_{cr}, deflections of the structure begin to grow. This growth of deflections can be limited or unlimited. Then, they noticed that for every moment of time t less than the critical time t_{cr}, the tangent stiffness matrix [30] was positively defined. The above allows for formulating the following eigenvalue problem:

$$(\mathbf{K_T} - \chi\mathbf{M})\vartheta = 0, \tag{5.1}$$

where $\mathbf{K_T}$ is the tangent stiffness matrix, $\mathbf{M}$ is the mass matrix of the structure, χ is an eigenvalue and ϑ is the vector of eigenvalues.

Analysing (5.1) for the time t equal to the critical value t_{cr}, it is found that the smallest eigenvalue $\chi_{min}(t_{cr})$ changes its sign from positive to negative, which corresponds to the growth of the structure deflection. The above analysis shows that the zero eigenvalue corresponds to singularity of the tangent stiffness matrix. In order to formulate the stability criterion, it should be determined whether an increase in the deflection at the time $t > t_{cr}$ is limited or unlimited. On the basis of the examples presented in [13], it has been noted that the smallest eigenvalue for the times $t > t_{cr}$ reaches the minimum value of χ'_{min} and then begins to rise and again changes its sign to the positive one and reaches the maximum value χ''_{min}.

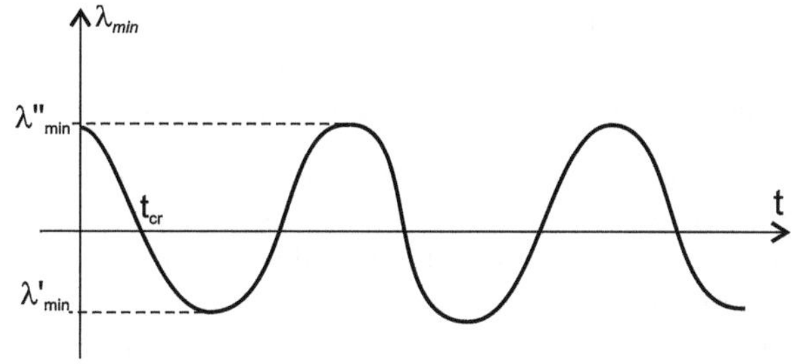

Fig. 5.6 Exemplary time history for the minimum eigenvalue

On this basis, Kleiber, Kotula and Saran defined a quasi-bifurcation criterion of dynamic stability for the structures subjected to pulse loading. It says that the structure loses its stability (deflection begins to grow indefinitely) when for $t = t_{cr}$ the determinant of the tangent stiffness matrix is equal to zero and the absolute value of the smallest eigenvalue χ'_{min} of (5.1) is greater than the maximum absolute value of the next smallest eigenvalue χ''_{min}(Fig. 5.6), i.e., $|\chi'_{min}| > \chi''_{min}$.

In summary, in order to consider the load as critical (causing an infinite deflection growth) for the given time $t = t_{cr}$, the two following conditions have to be simultaneously satisfied:

$$\mathbf{K_T} = 0 \wedge |\chi'_{\min}| > \chi''_{\min}. \tag{5.2}$$

The tangent stiffness matrix in the stability theory of dynamical systems [12] corresponds to the Jacobi matrix.

5.5 Author's Criterion

The above-discussed criteria can be applied to plates, shells or beam-columns and they have been formulated for the non-coupled buckling mode. In complex thin-walled structures, the stability loss has often a coupled form—one buckling mode enhances or accelerates the creation of another one and, as a result, a new buckling mode appears. Such multi-modal modes of the stability loss should also be taken into account in the analysis of thin-walled structures subjected to pulse load. As is well known from the dynamic buckling literature, the pulse duration equal to the period or half a period of natural vibrations with the mode corresponding to the buckling mode is considered.

Coupling of various buckling modes is associated with different vibration frequencies (and, thus, periods of vibration) corresponding to different buckling modes. This often leads to situations where for one buckling mode, the established pulse duration corresponds to the dynamic load and for another one, the period of vibration is so long that the pulse load should be treated as quasi-static. In the global buckling mode case, the deflection of the structure grows to infinity. Taking above into account, the following questions appear:

- how to determine the pulse duration in the case of interactive buckling?
- which value should be taken as the critical one in the global buckling mode—the deflection asymptote or the deflection for which the theory is no longer valid?

To answer the first question, numerous calculations and simulations have been performed. It has led to the conclusion that the worst case is when the pulse duration corresponds to the first flexural vibration mode—for short beam-columns it corresponds to the local mode. For longer structures, the first flexural vibration mode corresponds to the global mode, which finally appears when different buckling modes interact.

While answering the second question and taking the above-mentioned into account, one should note that the new dynamic buckling criterion could be formulated especially for all cases when the deflection grows rapidly to infinity. However, this new criterion could also work for the remaining buckling modes.

The equations of motion solution for the multi-mode (3.42) or single-mode (3.43) buckling allow for determining the deflection change of the analysed structure in time and, thus, the function that describes the behaviour of the system subjected to pulse load. Similar equations are solved in problems of stability of motion. Therefore, it has been decided to look for an analogy in determining the critical value describing dynamic buckling. The quasi-bifurcation criterion of dynamic stability for pulse load (Heaviside type) proposed by Kleiber, Kotula and Saran [13] seemed to be close to the dynamic buckling (impact) problem. The direct application of this criterion for estimation of the critical amplitude of pulse loading (of finite duration) for analysed thin-walled structures did not give expected results. The obtained dynamic critical load values were smaller than the results obtained from other well-known criteria [11, 28]. The discrepancies can be explained by different structures and different pulse types.

The further analysis concentrated on the Jacobi matrix calculation in a similar way as it is done in the dynamic stability problem for the periodical solution [12], where the values of the characteristic roots of the Jacobi matrix were checked. On the basis of numerous examples and many studies (the results are presented in Chap. 6.2), it was noticed that for the thin-walled structures subjected to pulse loading, which lose their stability according to the Budiansky-Hutchinson criterion or the Volmir criterion, the maximal radius $r_{\max}$ calculated from the characteristic root $\chi = a + jb$ (where $j = \sqrt{-1}$) of the Jacobi matrix was equal or greater than unity in the complex plane (Fig. 5.7). It was also noted that it is was sufficient to analyse the eigenvalues at any time between 0 and $1.5 \cdot T_p$. Therefore, the dynamic buckling criterion for thin-walled structures can be formulated as follows:

> *Thin-walled structures subjected to pulse loading of finite duration lose their stability even if one characteristic root $\chi = a + jb$ of the Jacobi matrix found for every time moment from 0 to $1.5T_p$ lies in the complex plane outside the circle with the radius equal to unity.*

The developed criterion has some limitation in application. The limitation was found during the dynamic buckling analysis of the plate subjected to rectangular pulse load [16]. For the mentioned case, it was observed that for pulse duration less

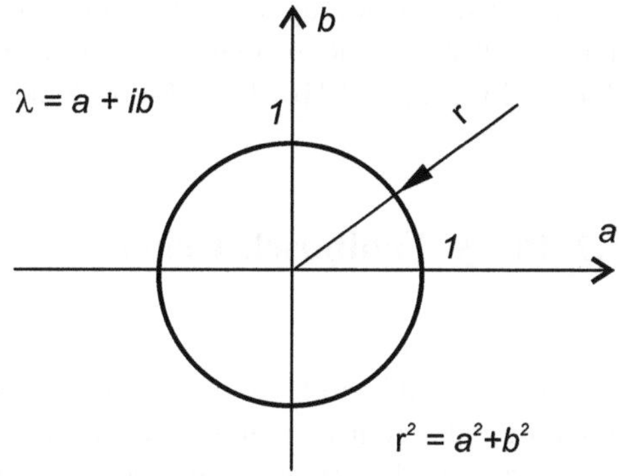

Fig. 5.7 Geometric representation of the proposed criterion

than $0.75T_p$ (T_p period of natural vibration), the results were independent of the pulse duration, thus for this case, it could be used only for pulse load with duration no longer than $0.75T_p$.

5.6 Teter Criteria

A similar way to find a dynamic buckling criterion as described in previous Section 5.5 was used by Teter [26]. Teter analysed the behaviour of thin-walled columns with stiffened open cross-sections subjected to pulse loading. The coupled buckling was analysed. As found out by Teter, the criterion proposed by the author of this monograph is not sufficient to estimate the critical amplitude of pulse load for long columns with stiffened open cross-sections [26], particularly if the duration of the pulse is equal to half a period of the vibration in the considered structure. Taking above into account, he attempted again to modify the Kotula-Kleiber-Saran criterion and formulated the following one:

A dynamic stability loss occurs when during the tracing time of solutions all eigenvalues of the Jacobi matrix are not positive simultaneously and at any moment one can find two negative eigenvalues.

The main disadvantage of this criterion is the same as in the previous one (Sect. 5.5)—a need to determine the Jacobi matrix (the tangent stiffness matrix) and its eigenvalues. This possibility exists for analytical-numerical methods or other open source software. Therefore, Teter decided to propose a new phase plane criterion. It is based on the results of numerical calculations obtained from any software to determine the displacement and the velocity. This criterion states:

The dynamic buckling load for the tracing time of solutions has been defined as the minimum value of pulse load such that the phase portrait is an open curve.

The criterion for dynamic buckling structures subjected to unbounded pulse load based on the phase plane was used by Schokker et al. [21], Hutchinson and Budiansky [11] and Hsu [10]. More details can be found in [3].

5.7 Petry–Fahlbusch Criterion

Petry and Fahlbusch [19] have noted that the Budiansky-Hutchinson criterion originally formulated for shells, widely used by many authors in the analysis of the dynamic behaviour of plates, does not allow one to use fully the capacity of the plate structure. In practice, this leads to a conservative determination of the critical dynamic load for the plate. These authors believe that the critical dynamic load should be based on the stress state analysis for the structures with a stable post-buckling equilibrium path. Analysing the stress state in any moment of time for the structures subjected to pulse load, it is possible to determine the dynamic load leading to a failure. On the basis of this approach, Petry and Fahlbusch have formulated the following criterion of dynamic buckling:

A dynamic response of the structure subjected to pulse load is dynamically stable if the condition that the equivalent stress is less than or equal to the assumed limit of stress is satisfied at any time and any point of the structure.

In the case when a deformable body is taken into consideration, they suggest assuming the yield limit as a limit of stress in the proposed criterion. Petry and Fahlbusch have redefined also the dynamic load factor to the following form:

$$DLF_f = \frac{N_F^{dyn}}{N_F^{stat}}, \tag{5.3}$$

where the dynamic failure load N_F^{dyn} and the static failure load N_F^{stat} in the sense of the limit stress are introduced. Petry and Fahlbusch in their work present results for the structures made of isotropic materials, in which, as is well known, the equivalent stress is calculated according to the Huber-Mises hypothesis. Nowadays, composite materials with orthotropic or even anisotropic material properties are very often used for different structures, so the Petry-Fahlbusch approach should be modified by an application of proper failure criteria for such materials.

5.8 A New Approach to Dynamic Buckling Load Estimation

Examining typical curves of the dynamic buckling problem (Fig. 5.8b) and the postbuckling behaviour of thin-walled structures subjected to static load (Fig. 5.8a), one can notice some similarity.

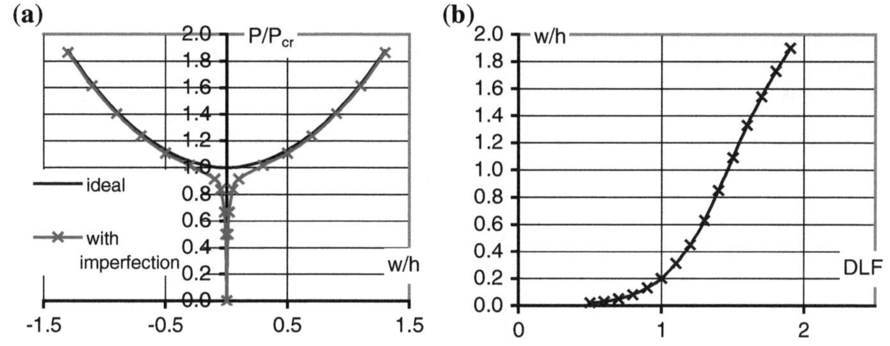

Fig. 5.8 Postbuckling equilibrium paths (**a**) and nondimensional displacement as a function of the dynamic load factor (**b**)

Fig. 5.9 Postbuckling behaviour in the case of static and dynamic load

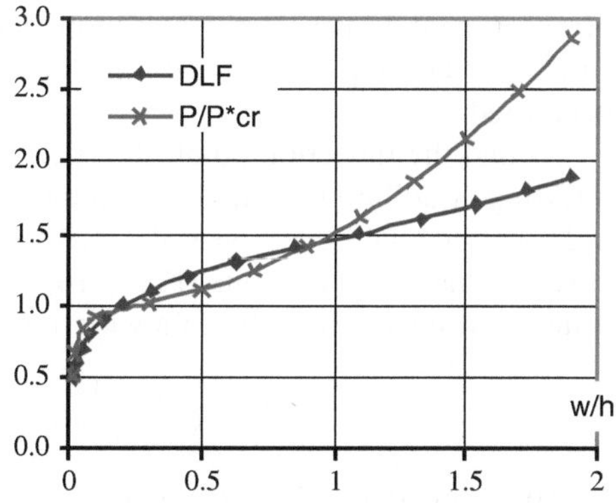

The curves shown in Fig. 5.9 (i.e., the nondimensional static load P/P^*_{cr} and the nondimensional dynamic load DLF as a function of nondimensional displacement) are similar, especially when the states are changed from prebuckling to post-buckling. Taking into account this similarity, it has been decided to check the suitability of the well-known methods for determining critical static loads, based on the results of experimental tests. The most popular methods are:

- Soutwell method [3, 18, 23, 24, 29].
- mean-strain (P-ε_m) method [7, 23, 24, 27];
- method of straight-lines intersection in the plot of mean strains [7, 23, 24, 27];
- alternative method P-w^2 [7, 23, 24, 27];
- P-w curve inflection point method [7];
- "top of the knee" method [15];
- Tereszkowski method [25];
- Koiter method [9, 14].

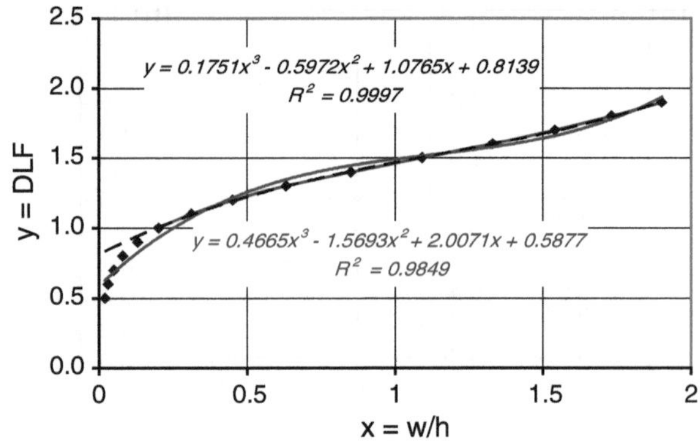

Fig. 5.10 Trend lines received from the results (points) of the numerical calculations

Kubiak and Kowal-Michalska [17] have decided to employ two well-known methods for identification of the critical load that are usually applied to the results of experimental investigations. The inflection point method (*P*-*w*), which is very similar to the "top of the knee" method, and the alternative (P-w^2) method have been used.

To find the inflection point, the approximation equations are found. The post-buckling equilibrium paths for structures with initial imperfections are the third order polynomial. It has been decided to adopt the same order function to fit the curve on the basis of points (Fig. 5.10) received from the numerical calculations. Thus, this function has the following form:

$$DLF = a_3\left(\frac{w}{h}\right)^3 + a_2\left(\frac{w}{h}\right)^2 + a_1\left(\frac{w}{h}\right) + a_0. \tag{5.4}$$

Approximating DLF(w/h) with function (5.4), it is very easy to find the inflection point and its coordinates:

$$\begin{aligned} \left(\frac{w}{h}\right)_{cr} &= \frac{-2a_2}{6a_3}, \\ DLF_{cr} &= a_3\left(\frac{w}{h}\right)^3_{cr} + a_2\left(\frac{w}{h}\right)^2_{cr} + a_1\left(\frac{w}{h}\right)_{cr} + a_0. \end{aligned} \tag{5.5}$$

As presented in Fig. 5.10, there are two different approximation curves (trend lines) and their equations—the first curve (the red one, bottom equation in Fig. 5.10) is obtained including all points from the numerical calculations and the second one (green, upper equation—Fig. 5.10) is based on points for DLF equal or higher than 1. The obtained results based on the above-mentioned curves are presented in Table 5.1.

Comparing the results presented in Table 5.1, it can be said that the critical values of DLF do not depend on the number of points taken into consideration for determining the trend line equation—it is enough to take all points for $DLF \geq 1$. More results obtained employing this approach and comparisons with other criteria will be presented in Sect. 6.2.

Table 5.1 An influence of taken approximate equations on critical value of DLF

Case	Constant in (5.4)				DLF_{cr}
	a_3	a_2	a_1	a_0	
1) All points from the numerical calculations taken to obtain an approximate equation	0.4665	−1.5693	2.0071	0.5877	1.523
2) Points with DLF ≥ 0.7	0.3252	−1.1071	1.5800	0.6832	1.526
3) Points with DLF ≥ 0.8	0.2655	−0.9081	1.3899	0.7291	1.527
4) Points with DLF ≥ 0.9	0.2159	−0.7396	1.2236	0.7722	1.527
5) Points with DLF ≥ 1.0	0.1751	−0.5972	1.0765	0.8139	1.523

To use the alternative (P-w^2) method, the relation between the dynamic load factor and dimensionless deflections should be changed into a curve or an equation describing the relation of *DLF* versus the square of the dimensionless deflection, and next, the point of intersection of the straight line representing the postbuckling state with the vertical axis can be found (Fig. 5.11). For exemplary results obtained for a simply supported compressed plate with the initial geometrical imperfection equal to 1/100 of the plate thickness (Fig. 5.11), the critical value of DLF_{cr} is equal to 1.14, according to the proposed alternative method (P-w^2).

The next suggestion concluded in [17] is to calculate the dynamic load factor as a relation between the amplitude of the pulse load and the critical load for an imperfect structure. According to the obtained results [17] (also presented in Sect. 6.2), it can be said that for small imperfections (i.e., less than or equal to 1/100 of the plate or wall thickness), there are no differences in the results after assuming the traditional definition of the dynamic load factor DLF. However, for higher imperfections (i.e., greater than 1/10 of the plate thickness), the relations between the nondimensional deflection and the dynamic load factor are similar to static postbuckling equilibrium paths, which could mean that for highly imperfect structures the dynamic responses are similar to deflection for statically loaded structures. More examples are presented in Sect. 6.2.

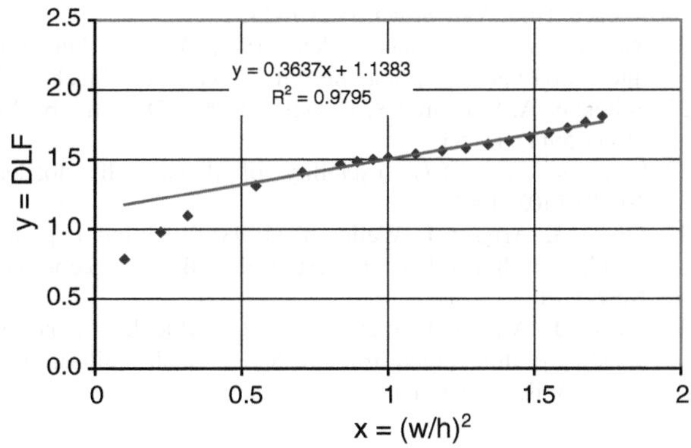

Fig. 5.11 Linear trend line for the alternative method

References

1. Abramovich H, Grunwald A (1995) Stability of axially impacted composite plates. Compos Struct 32:151–158
2. Ari-Gur J, Simonetta SR (1997) Dynamic pulse buckling of rectangular composite plates. Compos B 28B:301–308
3. Bazant ZP, Cedolin L (1991) Stability of structures: elastic, inelastic, fracture and damage theories. Oxford University Press, New York
4. Budiansky B (1965) Dynamic buckling of elastic structures: criteria and estimates. Report SM-7, NASA CR-66072
5. Budiansky B, Roth RS (1962) Axisymmetric dynamic buckling of clamped shallow spherical shells. Collected Papers on Instability of Shell Structures, NASA, TN-D-1510:597–606
6. Cheong HK, Hao H, Cui S (2000) Experimental investigation of dynamic post-buckling characteristics of rectangular plates under fluid-solid slamming. Eng Struct 22:947–960
7. Coan JM (1951) Large-deflection theory for plates with small initial curvature loaded in edge compression. ASME J Appl Mech 18:143–151
8. Cui S, Hao H, Cheong HK (2001) Dynamic buckling and post buckling of imperfect columns under fluid-solid interaction. Int J Solids Struct 38:8879–8897
9. van der Heijden AMA (ed) (2009) W.T. Koiter's Elastic Stability of Solids and Structures. Cambridge University Press, New York
10. Hsu CS (1967) The effects of various parameters on the dynamic stability of shallow arch. J Appl Mech 34(2):349–356
11. Hutchinson JW, Budiansky B (1966) Dynamic buckling estimates. AIAA J 4–3:525–530
12. Kapitaniak T, Wojewoda J (2000) Bifurkacje i chaos. PWN, Warsaw-Lodz
13. Kleiber M, Kotula W, Saran M (1987) Numerical analysis of dynamic quasi-bifurcation. Eng Comput 4(1):48–52
14. Koiter WT (1963) Elastic stability and post-buckling behaviour. In: Proceedings of the symposium on non-linear problems, University of Wisconsin Press, Wisconsin, pp 257–275
15. Kolakowski Z, Kowal-Michalska K (eds) (2012) Static, dynamic and stability of structures. Vol. 2: Statics, dynamics and stability of structural elements and systems, A series of monographs. Technical University of Lodz Press, Lodz
16. Kubiak T (2007) Interakcyjne wyboczenie dynamiczne cienkosciennych slupow. Technical University of Lodz Press, Lodz
17. Kubiak T, Kowal-Michalska K (2012) A new approach to dynamic buckling load estimation for plate structures. In: Proceedings of Stability of Structures 13-th Symposium, Zakopane, Poland, pp 397–406
18. Parlapalli MR, Soh KC, Shu DW, Ma G (2007) Experimental investigation of delamination buckling of stitched composite laminates. Compos A 38:2024–2033
19. Petry D, Fahlbusch G (2000) Dynamic buckling of thin isotropic plates subjected to in-plane impact. Thin Wall Struct 38:267–283
20. Raftoyiannis IG, Kounadis AN (2000) Dynamic buckling of 2-DOF systems with mode interaction under step loading. Int J Non-Linear Mech 35:531–542
21. Schokker A, Sridharan S, Kasagi A (1996) Dynamic buckling of composite shells. Comput Struct 59(1):43–55
22. Simitses GJ (1987) Instability of dynamically loaded structures. Appl Mech Rev 40(10):1403–1408
23. Singer J, Arbocz J, Weller T (1998) Buckling experiments. Experimental methods in buckling of thin-walled structure. Vol. 1. Basic concepts, columns, beams, and plates. Wiley, New York
24. Singer J, Arbocz J, Weller T (2002) Buckling experiments. Experimental methods in buckling of thin-walled structure. Vol. 2. Shells built-up structures, composites and additional topics. Wiley, New York

25. Tereszkowski Z (1970) An experimental method for determining critical loads of plates. Arch Mech Eng 3:485–493
26. Teter A (2011) Dynamic critical load based on different stability criteria for coupled buckling of columns with stiffened open cross-sections. Thin Wall Struct 49:589–595
27. Venkataramaiah KR, Roorda J (1982) Analysis of local plate buckling experimental data. In: Proceedings of 6th international specialty conference on cold-formed steel structures, Missouri S&T, pp 45–74
28. Volmir SA (1972) Nieliniejnaja dinamika plastinok i oboloczek. Science, Moscow
29. Wong PMH, Wang YC (2007) An experimental study of pultruded glass fibre reinforced plastics channel columns at elevated temperatures. Compos Struct 81:84–95
30. Zienkiewicz OC, Taylor RL (2000) The finite element method. Butterworth-Heinemann, Oxford

Chapter 6
Thin Plates

The chapter is devoted to buckling, postbuckling behaviour and dynamic buckling of thin plates made of metals or composites modelled as orthotropic materials. The considered plates are simply supported on loaded edges and different boundary condition on longitudinal edges. They are subjected to uniform compression.

All calculations were performed assuming elastic homogenous material properties. Isotropic and orthotropic materials were considered (Table 6.1).

All materials mentioned in Table 6.1 are isotropic, hence, the Kirchhoff's modulus can be calculated according to following well known equation:

$$G = \frac{E}{2(1-\nu)}. \tag{6.1}$$

The fibre composite material was modelled as orthotropic but for components (resin and fibre), isotropic material properties (Table 6.1) were assumed. Necessary equations for material properties homogenization based on the theory of mixture [2, 3] are as follows:

$$\begin{aligned} E_x &= E_m(1-f) + E_f f, \\ E_y &= E_m \frac{E_m(1-\sqrt{f}) + E_f\sqrt{f}}{E_m[1-\sqrt{f}(1-\sqrt{f})] + E_f\sqrt{f}(1-\sqrt{f})}, \\ \nu_{yx} &= \nu_m\left(1-\sqrt{f}\right) + \nu_f\sqrt{f}, \\ G &= G_m \frac{G_m\sqrt{f}(1-\sqrt{f}) + G_f[1-\sqrt{f}(1-\sqrt{f})]}{G_m\sqrt{f} + G_f(1-\sqrt{f})}, \end{aligned} \tag{6.2}$$

where E_m and E_f are the Young's modulus of elasticity for matrix and fibre, respectively, G_m and G_f are the shear modulus for matrix (subscript *m*) and fibre (subscript *f*), ν_m and ν_f are Poisson's ratios for matrix and fibre, and $f = V_f/(V_m + V_f)$ is the fibre volume fraction.

It should be noticed that taking into account (6.2) and the volume fibre fraction f from the range 0.2 to 0.7, the coefficient of orthotropy defined as E_x/E_y varies from 2.9 to 3.6. Moreover, in real, industrially produced structures, the

T. Kubiak, *Static and Dynamic Buckling of Thin-Walled Plate Structures*,
DOI: 10.1007/978-3-319-00654-3_6,

Table 6.1 Assumed material properties

Material type	E (GPa)	ν	ρ (kg/m^3)
Steel	200	0.3	7,850
Aluminium	70	0.33	2,950
Epoxy resin	3.5	0.33	1,249
Glass fibre	71	0.22	2,450

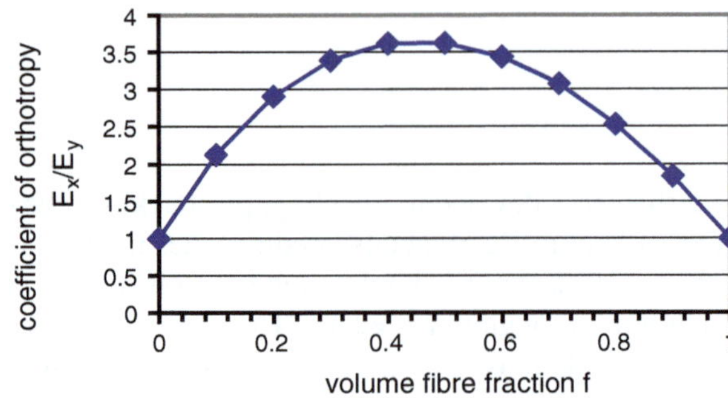

Fig. 6.1 Volume fibre fraction influence on coefficients of orthotropy for fibre glass epoxy resin

above-mentioned volume fibre fraction f ranges from 0.4 to 0.6, so in our case the coefficient of orthotropy varies from 3.4 to 3.6. Figure 6.1 shows a relation between the volume fibre fraction and the coefficient of orthotropy for glass fibre in epoxy resin with the material data presented in Table 8.1.

Rectangular thin plates simply supported on loaded edges with different boundary conditions along the unloaded edges were considered (Fig. 6.2). On the longitudinal edges, five different boundary condition cases were taken into account. The following notations are used in Fig. 6.2: s—simply supported edge, c—clamped edge, e—free edge.

Plates with constant and variable material properties were considered. Composite plates with the widthwise variable volume fibre fraction were modelled by dividing the plate into strips (Fig. 6.3) with constant material properties assigned to the assumed volume fibre fraction f. Its value was determined on the basis of the arbitrary adopted sine function:

$$f = f_{av} + A \cdot \cos\left(\frac{2\pi y}{b}\right), \tag{6.3}$$

where: $f_{av} = 0.5$—arbitrary assumed average value of the fibre volume fraction; $A = < -0.4; 0.4 >$—amplitude of sine describing the change of material properties along the plate width. The range of the amplitude is assumed in such a way that the volume fibre fraction f varies from 0.1 to 0.9 or from about 10 % to 90 % of reinforcing fibres in the composite structure. However, the proposed method allows one to analyse any function describing a widthwise variation of material properties, not necessarily that one defined by changes in the volume fibre fraction content f [6, 7].

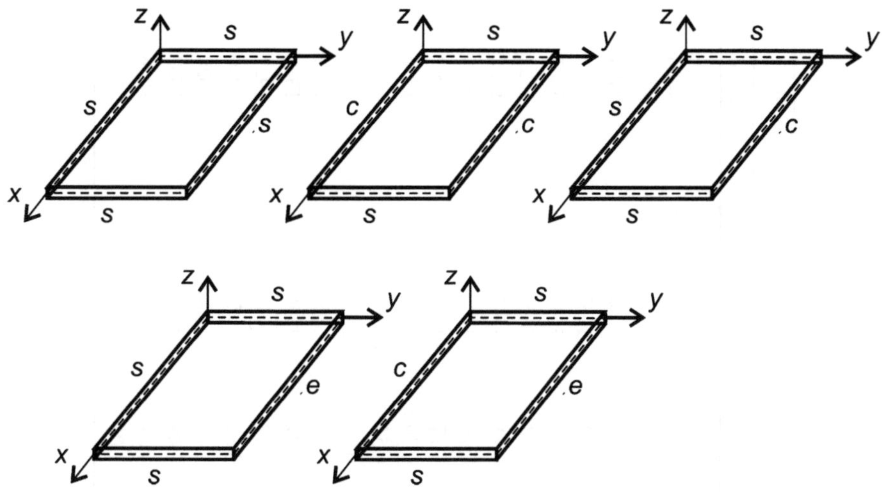

Fig. 6.2 Analysed plates with different boundary conditions

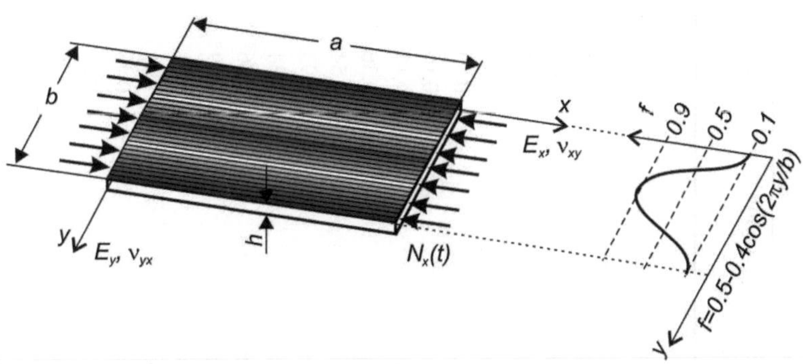

Fig. 6.3 Model of the plate with widthwise sinusoidally variable material properties (volume fibre fraction)

The results presented further were obtained using the analytical-numerical method (MAN) and the finite element method (FEM) software.

In the dynamic buckling analysis, load is defined as a dynamic load factor, i.e., an amplitude of the pulse load divided by the static buckling load. Therefore, the static buckling load and the corresponding buckling mode should be determined. The buckling mode is used to map an initial imperfection on the plate midplane. The time of pulse duration is assumed as corresponding to a period of natural vibrations. It means that the eigenvalue modal analysis and the eigenbuckling analysis should be performed in the first stage.

The natural frequencies and the buckling load for composite and steel square plates ($a/b = 1$) with the thickness ratio $b/h = 100$ for different boundary condition cases are presented in Table 6.2.

Table 6.2 Buckling load and natural frequencies for the plates under analysis

Material	Boundary conditions	Buckling load P_{cr} (kN)		Natural frequencies ω (rad/s)	
		MAN	FEM	MAN	FEM
Steel	ss	7.23	7.24	3016	3010
	cc	15.55 ($m = 1$)	15.66	4423 ($m = 1$)	4423
		13.90 ($m = 2$)	14.02	8363 ($m = 2$)	8344
	se	2.53	2.54	1784	1784
	ce	2.99	2.99	1935	1935
	sc	10.38	10.41	3613	3607
Glass fibre epoxy resin composite with $f = 0.2$	ss	0.28	0.28	1351	1351
	cc	0.50	0.51	1822	1822
	se	0.16	0.16	1030	1030
	ce	0.17	0.17	1062	1062
	sc	0.36	0.36	1546	1546
Glass fibre epoxy resin composite with $f = 0.5$	ss	0.54	0.54	1703	1709
	cc	0.93	0.95	2231	2237
	se	0.34	0.35	1351	1351
	ce	0.36	0.37	1389	1389
	sc	0.69	0.70	1916	1923
Glass fibre epoxy resin composite with $f = 0.7$	ss	0.81	0.82	1954	1960
	cc	1.44	1.46	2608	2620
	se	0.48	0.48	1502	1502
	ce	0.51	0.51	1558	1558
	sc	1.05	1.06	2224	2231

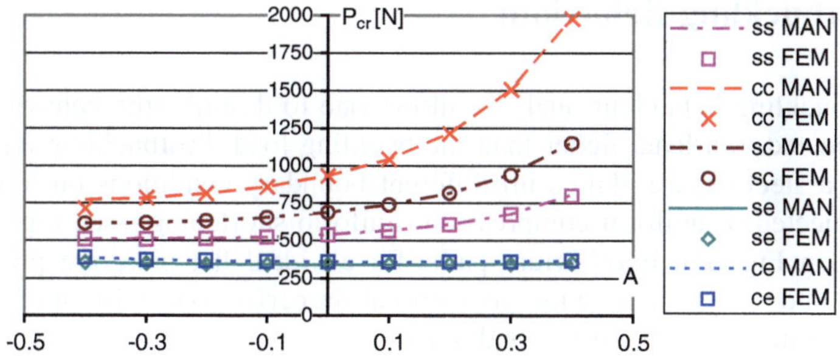

Fig. 6.4 Static buckling load P_{cr} as a function of the amplitude A describing the distribution of the volume fibre fraction

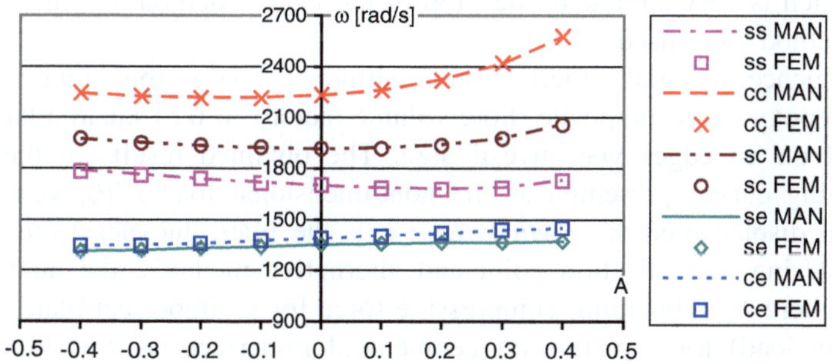

Fig. 6.5 Natural frequencies as a function of the amplitude A describing the distribution of the volume fibre fraction

The natural frequencies and the buckling load for the square composite plate (a/b = 1, b/h = 100) with the widthwise variable volume fibre fraction (6.3) for different amplitudes A (describing the volume fibre fraction distribution) are presented in Figs. 6.4 and 6.5, respectively. For all boundary condition cases, both the employed methods of calculations give similar results, which confirms the correctness of the calculations.

As can be seen in Figs. 6.4 and 6.5, the buckling loads as well as the natural frequencies grow with an increasing value of the amplitude A describing the volume fibre fraction in the plate widthwise direction. This means that plates with stiffer longitudinal strips near the plate edges are stiffer as a whole (a higher buckling load and higher natural frequencies) than plates with a stiffer central longitudinal strip.

6.1 Postbuckling Behaviour

The postbuckling behaviour analysis allows one to describe the behaviour of the plate subjected to a load higher than the buckling load. Postbuckling equilibrium paths for a steel square plate with different boundary conditions on longitudinal edges subjected to uniform compression (uniform shortening—see Fig. 6.11) are presented in Fig. 6.6. Equilibrium paths for an ideal flat plate are presented in Fig. 6.6a and for a plate with geometrical imperfections with the amplitude $\xi^* = 0.01$—in Fig. 6.6b, correspondingly.

For an ideal plate structure, the critical load can be determined from the eigenvalue analysis but for structures with imperfections, the buckling load may be determined on the basis of the pre- and post- buckling behaviour. Two well-known methods for identification of the critical load were employed. They are usually applied to the results of experimental investigations. The inflection point method (P-w), which is very similar to the "top of the knee" method and the alternative (P-w^2) method were used.

An influence of initial imperfection amplitudes ξ^* on postbuckling equilibrium paths for epoxy glass composite (fibre volume factor $f = 0.5$) square plates simply supported on all edges was investigated. The obtained results in the form of postbuckling curves presented as the nondimensional load P/P_{cr} versus nondimensional displacement $\xi = w/h$ (where h is the plate thickness) are shown in Fig. 6.7. Using the inflection point and alternative methods, the buckling load P^*_{cr}/P_{cr} (where P^*_{cr}—buckling compressive force for an imperfect plate and P_{cr}—bifurcation load) for a plate with geometrical imperfections was found and is presented in Table 6.3.

The results shown in Table 6.3 allow one to conclude that lower values of P^*_{cr} were obtained using the P-w^2 method and the differences between the results of both the methods grow with an increase in the imperfection amplitude value. It can also be noted that the higher initial imperfection amplitude, the lower critical buckling load.

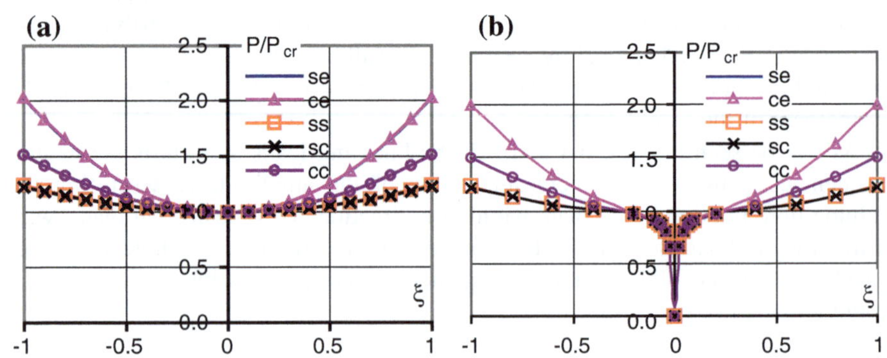

Fig. 6.6 Postbuckling equilibrium paths for square ideal plates (**a**) and plates with imperfections (**b**) with different boundary conditions on non-loaded edges

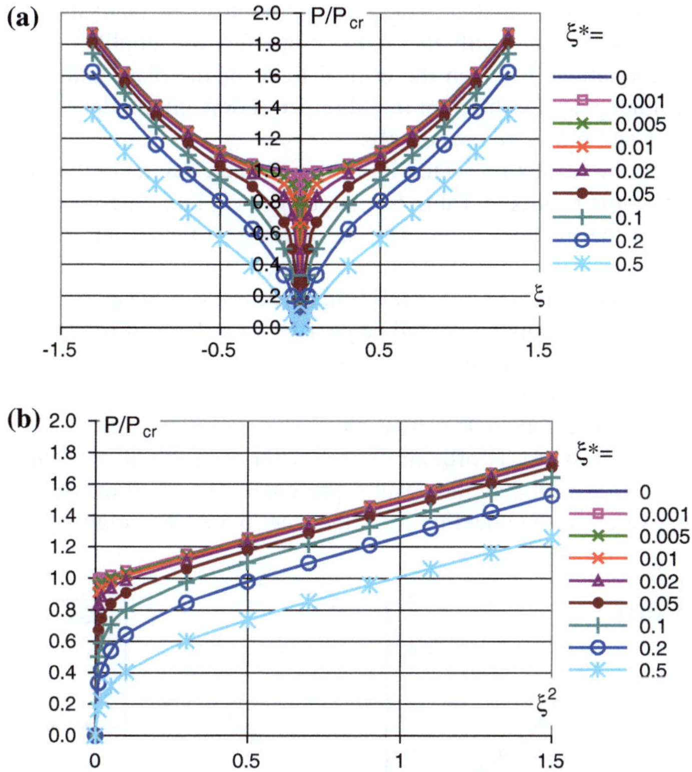

Fig. 6.7 Dimensionless load P/P$_{cr}$ versus dimensionless deflection (**a**) or square of dimensionless deflection (**b**)

Table 6.3 P^*_{cr}/P_{cr} for different amplitudes of initial imperfections

Determination method: initial imperfection amplitude ξ^*	P-w P^*_{cr}/P_{cr}	P-w^2 P^*_{cr}/P_{cr}
0.001	0.999	0.999
0.005	0.998	0.994
0.01	0.995	0.986
0.02	0.990	0.968
0.05	0.968	0.925
0.1	0.925	0.863
0.2	0.834	0.751
0.5	0.589	0.494

An influence of the initial imperfection amplitude on the critical buckling load and the postbuckling behaviour was checked for the rest of the assumed boundary conditions (Fig. 6.2). Some exemplary postbuckling equilibrium paths obtained with MAN for rectangular composite (volume fibre fraction $f = 0.5$) plates are presented in Figs. 6.8 and 6.9. The results for an ideal plate are presented in Fig. 6.8 and for a plate with the initial imperfection amplitude $\xi^* = 0.1$—in Fig. 6.9. The relations between the critical load for the plate with imperfections and the ideal flat plate for different initial imperfection amplitudes ξ^* are presented in Tables 6.4 and 6.5, correspondingly.

Comparing the results presented in Tables 6.4 and 6.5, a similar conclusion to those based on the results in Table 6.3 can be drawn, i.e., differences between the results of both the methods grow with an increase in the imperfection amplitude value and the higher initial imperfection amplitude, the lower critical buckling load.

Let us compare postbuckling equilibrium paths for a plate made of steel and composite with all simply supported edges (the case denoted by "ss" in Fig. 6.2) and with one free edge (the case denoted by "se" in Fig. 6.2). A comparison of the results is presented in Fig. 6.10.

The results presented in Fig. 6.10 show that the plate made of a composite material is stiffer due to its orthotropic character than the plate made of steel. It should be emphasized that the results presented are nondimensional so for a dimensional value of the buckling load, the steel plate has a higher value than the

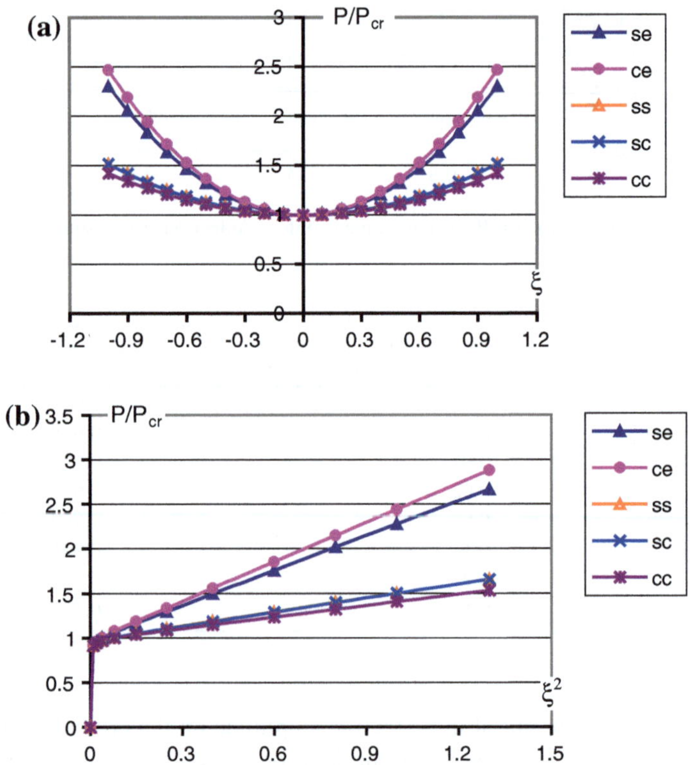

Fig. 6.8 Postbuckling equilibrium paths for ideal flat composite plates obtained from P-w (**a**) and P-w^2 (**b**) methods

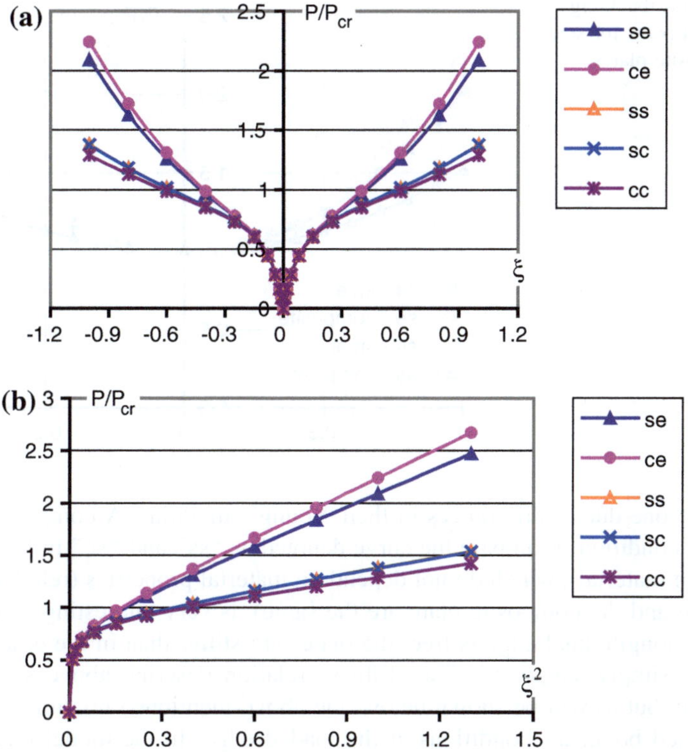

Fig. 6.9 Postbuckling equilibrium paths for plates with imperfections $\xi^* = 0.1$ obtained from P-w (**a**) and P-w^2 (**b**) methods

Table 6.4 P^*_{cr}/P_{cr} determined with the "top of the knee" method—P-w method

ξ^* = boundary conditions	0.01	0.1	0.5
se	0.99	0.87	0.42
ce	0.99	0.86	0.39
ss	1	0.93	0.59
sc	1	0.93	0.59
cc	1	0.93	0.62

Table 6.5 P^*_{cr}/P_{cr} determined with the alternative method—P-w^2 method

ξ^* = boundary conditions	0.01	0.1	0.5
se	0.97	0.77	0.38
ce	0.97	0.76	0.37
ss	0.97	0.79	0.43
sc	0.97	0.79	0.43
cc	0.97	0.79	0.44

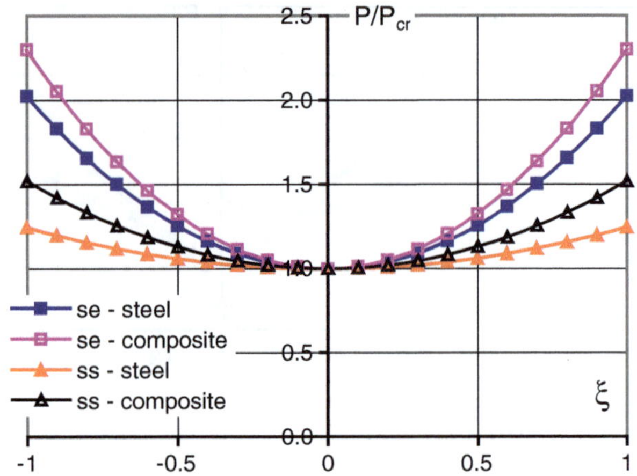

Fig. 6.10 Postbuckling equilibrium paths for steel and composite plates

composite one due to differences in their Young's modulus. A comparison of the boundary conditions (compare the curve denoted as "ss" and "se" in Fig. 6.10 for the chosen material), which do not depend on material properties (relations for the steel plate and the composite plate are the same), is very interesting—in the case when one longitudinal edge is free, the plates are stiffer than in the case when all edges are simply supported. The stiffness relation describe above seems to be unrealistic, but as will be shown below, the above-mentioned relation depends on the assumed boundary conditions at the loaded edge. In the model under investigation, the boundary conditions on loaded edges were assumed in such a way that the edges were straight and remained parallel during loading (Fig. 6.11).

A comparison between compressed simply supported plates with one longitudinal edge free for two different assumptions can be found in [5] and is presented in Fig. 6.12. The curve denoted by '1' was obtained on the assumption that the loaded edges were straight and remained parallel during loading. The curve denoted as '2' was obtained on the assumption that the loaded edges were straight and could rotate about normal to the middle surface plane. Comparing the postbuckling equilibrium paths denoted as 1 and 2 in Fig. 6.11, an influence of the assumed boundary conditions is very well visible. Additionally, in Fig. 6.11, a postbuckling path for the plate with all simply supported edges is presented and denoted as '3'.

6.2 Dynamic Buckling

Having determined the natural vibration frequency, the static buckling load with the corresponding buckling mode and postbuckling equilibrium paths for plates with initial geometrical imperfections, a dynamic response analysis can be performed for

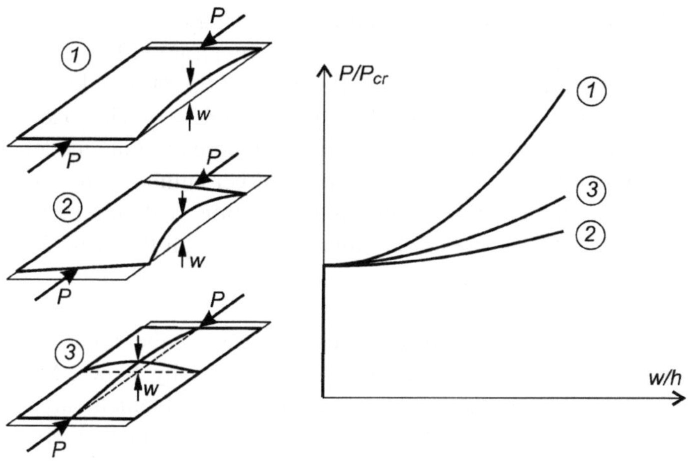

Fig. 6.11 Postbuckling equilibrium paths for plates with different assumptions on loaded edges

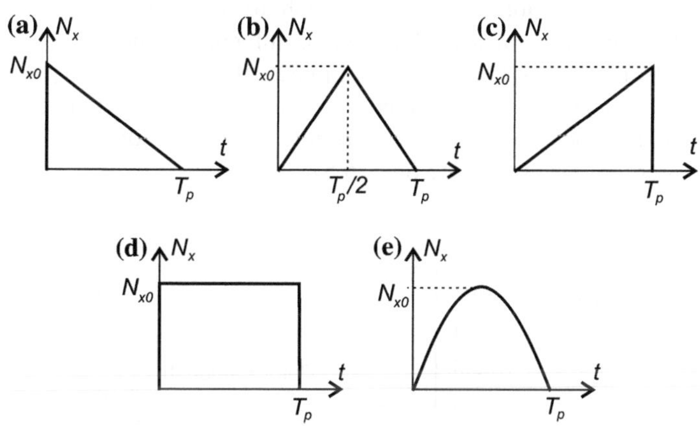

Fig. 6.12 Pulse load shapes denoted as: **a** T1, **b** T2, **c** T3, **d** P, **e** S

plates subjected to the pulse load (Fig. 6.12) of a rectangular (P), three triangular type (descending T1, equilateral T2 and T3 growing) and sinusoidal (S) shape.

In order to verify the analytical-numerical method, the finite element analysis was performed and the obtained results were compared to the results presented by Petry and Fahlbusch [12]. Figure 6.13 shows a dimensionless maximum deflection as a function of the dynamic load factor *DLF* for the aluminium square plate ($a/b = 1$ and $b/h = 200$) simply supported on all edges and subjected to compression with a sinusoidally shaped pulse load. The duration of the pulse corresponds to the period of natural vibrations for the plate under consideration. The amplitude of initial imperfections was assumed to be equal to 5/100 of the plate thickness.

The analytical-numerical method presented in Chap. 3 gives a slightly higher deflection value than the results presented in the literature [12] and obtained from the finite element method. The higher differences of the *DLF* value increase from about 2 % for the $DLF = 1.2$ up to 14 % for the $DLF = 4$. The discrepancies for the growing pulse load amplitude may be due to a more accurate model adopted by Petry and Fahlbusch [12], who took a relatively larger number of terms in the series function describing the plate deflection in the solution. It should be noted that the proposed approach allows one to determine well enough the critical value of the dynamic load factor DLF_{cr} according to the Volmir or Budiansky-Hutchinson criterion (see Table 6.6).

Other examples which confirm the correctness of the applied methods of computations can be found in the monograph edited by Kowal-Michalska [4]. Therefore, in the following part of the study, the results obtained only with the two employed methods of calculations, i.e., the finite element method (FEM) and the analytical-numerical method (MAN), will be compared.

Dynamic buckling of thin plates, as shown in the literature overview (Sect. 1.3.2), is the subject of many papers and has appeared in the literature for more than 50 years. The author of this monograph in his previous works [4, 8–10] analysed different shapes of pulse loading, an influence of material properties and boundary conditions on unloaded edges of plates. Below, a summary of the results contained in those works is presented. The results are compared to the results obtained with methods known from experimental investigations, i.e., the inflection point and alternative method (see Sect. 5.8).

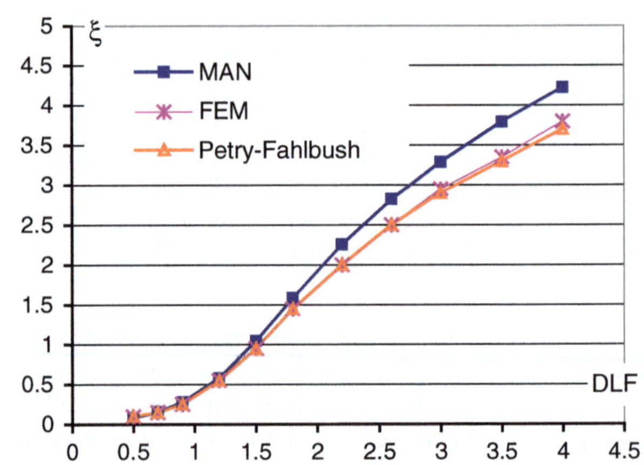

Fig. 6.13 Curves ξ (DLF) for the aluminium plate—a comparison of the results

Table 6.6 Critical value of *DLF* obtained with different methods

Criterion/method	Budiansky-Hutchinson	Volmir $\xi = 1$
MAN	1.5–1.8	1.48
FEM	1.5–1.8	1.53
Petry-Fahlbusch	1.5–1.8	1.52

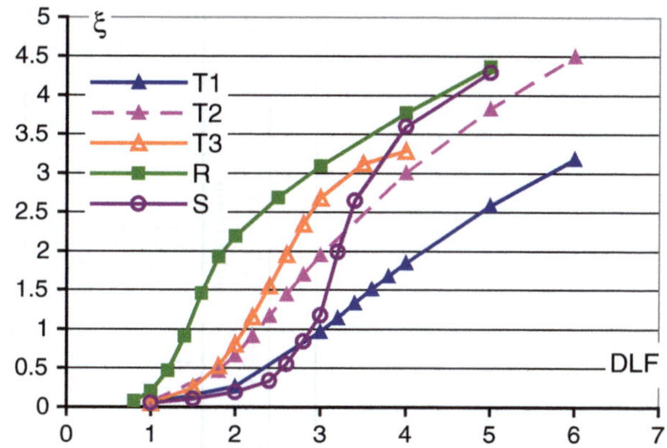

Fig. 6.14 Dynamic response curves for the simply supported plate—a comparison of the pulse load shape

A different shape of the pulse loading influence on dynamic responses of the plate is shown in Figs. 6.14 and 6.15 and in Table 6.7. Figure 6.14 presents some exemplary results obtained from the analytical-numerical method for the plate made of a composite with the volume fibre fraction equal to $f = 0.5$ and clamped on longitudinal edges ('cc'). In Fig. 6.15, a comparison of the results obtained with the analytical-numerical method (MAN) and the finite element method (*FEM*) for triangularly (*T1*) and rectangularly (*R*) shaped pulses is presented. The curves in Fig. 6.15 present the results for the simply supported plate made of an epoxy-glass composite with the volume fibre fraction equal to $f = 0.5$. The results obtained from both the methods are consistent.

The Budiansky-Hutchinson and Volmir criteria were compared for a square plate with the fibre volume fraction $f = 0.8$, for all the boundary conditions and impulses under analysis. The obtained critical value of dynamic load factors DLF_{cr} are presented in Table 6.7.

The dimensionless critical value of the dynamic load factor DLF_{cr} determined according to the Budiansky-Hutchinson criterion, the Volmir criterion, author's modification of the Kleiber-Kotula-Saran criterion and the methods known from

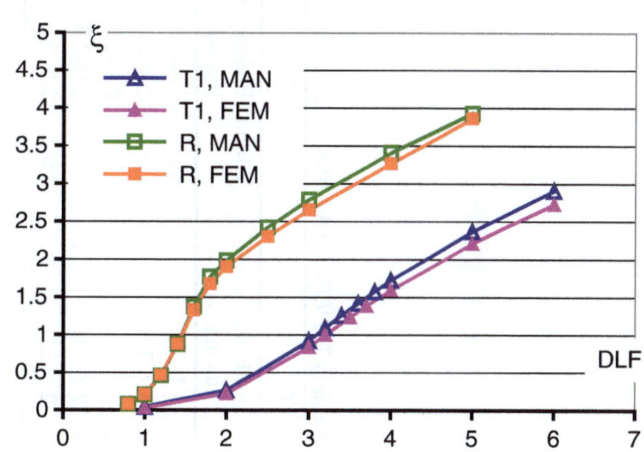

Fig. 6.15 Dynamic response curves for the simply supported plate—a comparison of calculation methods

Table 6.7 *DLF*s for the fibre volume fraction equal to f = 0.8

	T1		T2		T3		R		S	
	Budiansky-Hutchinson	Volmir	Budiansky-Hutchinson	Volmir	Budiansky-Hutchinson	Volmir	Budiansky-Hutchinson	Volmir	Budiansky-Hutchinson	Volmir
ss	3.1–3.3	3.1	2.4–2.6	2.3	2.4–2.6	2.1	1.4–1.6	1.4	3.0–3.2	2.9
cc	3.2–3.4	3.0	2.4–2.6	2.2	2.5–2.7	2.1	1.4–1.6	1.4	3.0–3.2	2.9
cs	3.2–3.4	3.1	2.4–2.6	2.3	2.4–2.6	2.1	1.4–1.6	1.4	3.0–3.2	2.9
se	2.8–3.0	3.6	2.0–2.2	2.6	2.0–2.2	2.2	1.2–1.4	1.6	3.0–3.2	3.0
ce	2.8–3.0	3.6	2.0–2.2	2.6	2.0–2.2	2.2	1.2–1.4	1.7	3.0–3.2	3.1

experimental studies (i.e., inflection point and alternative methods) for steel and composite plates subjected to a rectangularly shaped pulse load (with a duration corresponding to the period of natural vibrations) are summarized in Tables 6.8 and 6.9. The results were obtained with the analytical-numerical method on the assumption that the amplitude of initial geometrical imperfections was equal to $\xi^* = 0.01$.

The dimensionless critical dynamic load factor values presented in Tables 6.8 and 6.9 confirm a compliance of the new criterion and the new approach with the well-known Budiansky-Hutchinson and Volmir criteria. However, the alternative method used to determine the critical dynamic buckling amplitude yields the results about twice lower than the other applied methods. This means that the alternative method should rather not be used to determine the critical value of *DLF*, so in the further part of the presentation of the results, it will not be used. In the case of plates with boundary conditions denoted as "se" and "ce", the results

Table 6.8 DLF_{cr} comparison for steel plates

Boundary condition	Mode m	Critical value of the dynamic load factor DLF_{cr}				
		Budiansky-Hutchinson	Volmir $\xi = 1$	Author's criterion $r_{max} = 1$	P-w	P-w^2
ss	1	1.5–1.6	1.43	1.51	1.67	0.76
cc	2	1.4–1.5	1.51	1.38	1.50	0.69
se	1	1.5–1.6	1.55	1.35	1.22	0.59
ce	1	1.3–1.4	1.58	1.35	1.22	0.59
sc	1	1.4–1.8	1.46	1.58	1.70	0.76

Table 6.9 DLF_{cr} comparison for epoxy-glass composite plates

Volume fibre fraction f	Boundary condition	Mode m	Critical value of the dynamic load factor DLF_{cr}				
			Budiansky-Hutchinson	Volmir $\xi = 1$	Author's criterion $r_{max} = 1$	P-w	P-w^2
0.2	ss	1	1.45–1.6	1.46	1.46	1.52	0.69
	cc	1	1.45–1.6	1.44	1.49	1.53	0.72
	se	1	1.3–1.45	1.65	1.31	1.15	0.55
	ce	1	1.3–1.45	1.68	1.30	1.10	0.52
	sc	1	1.6–1.75	1.45	1.46	1.52	0.70
0.5	ss	1	1.45–1.5	1.46	1.45	1.45	0.69
	cc	1	1.45–1.6	1.45	1.48	1.52	0.71
	se	1	1.3–1.45	1.64	1.31	1.14	0.54
	ce	1	1.3–1.45	1.71	1.30	1.10	0.52
	sc	1	1.3–1.45	1.46	1.44	1.45	0.69
0.7	ss	1	1.3–1.45	1.34	1.32	1.43	0.66
	cc	1	1.45–1.6	1.44	1.49	1.53	0.72
	se	1	1.3–1.45	1.61	1.31	1.15	0.55
	ce	1	1.3–1.45	1.70	1.30	1.10	0.52
	sc	1	1.45–1.6	1.46	1.46	1.46	0.70

obtained with author's criterion differ by about 10 % and by 15 % when the inflection point method is used. In other cases (excluding the alternative method), they are between the values of the criteria designated by Budiansky-Hutchinson and Volmir. The best agreement of the results was achieved for plates with supported edges (i.e., the boundary condition denoted as 'ss', 'sc', 'cc') between the following two pairs: author's criterion—the Volmir criterion and the Budiansky-Hutchinson criterion—the inflection point method.

In the literature [1, 13–15], the dynamic buckling occurs when the pulse duration is close to a period of fundamental vibrations or to half a period of fundamental vibrations [1] and the initial deflection is very small in relation to the thickness of the plate. Therefore, the simply supported composite plate made of an epoxy-glass composite with the volume fibre fraction $f = 0.5$ subjected to rectangular pulse load was taken as an example to analyse an influence of pulse duration and the assumed initial imperfection amplitude on the dynamic response. The results are shown in Figs. 6.16 and 6.17 and are summarized in Tables 6.10 and 6.11, respectively.

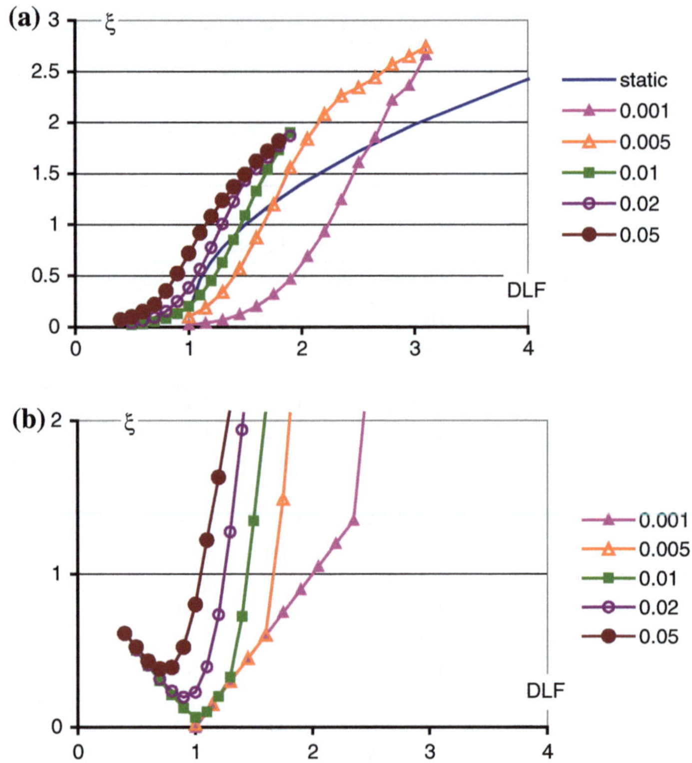

Fig. 6.16 Influence of the initial imperfection amplitude on the course of the curves ξ (*DLF*) and r_{max} (*DLF*) for the simply supported composite plate

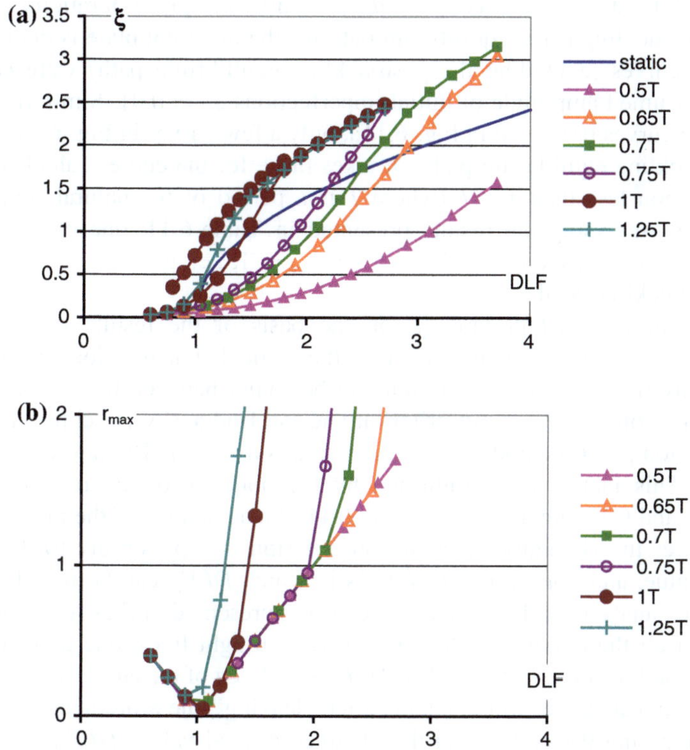

Fig. 6.17 Influence of the pulse duration on the course of the curves ξ (*DLF*) and $r_{\max}$ (*DLF*) for the simply supported composite ($f = 0.5$) plate

In Fig. 6.16, a postbuckling static equilibrium path for the ideal flat plate is also presented. The curves ξ (*DLF*) for a large amplitude of initial imperfections do not cross the postbuckling equilibrium path, which would suggest that for such large initial imperfections, the critical dynamic load factor is equal or even less than the static buckling load. However, the results summarized in Table 6.10 show that this is not entirely true and a decrease in DLF_{cr} with an increase in the amplitude of initial imperfections is observed.

Table 6.10 Influence of the amplitude of initial imperfections on DLF_{cr}

Amplitude of initial imperfections ξ^*	Critical value of the dynamic load factor DLF_{cr}			
	Budiansky-Hutchinson	Volmir $\xi = 1$	Author's criterion $r_{max} = 1$	Inflection point method *P-w*
0.001	2.35–2.5	2.23	2.00	2.40
0.005	1.75–1.9	1.66	1.67	1.68
0.01	1.45–1.5	1.46	1.45	1.52
0.02	1.2–1.3	1.30	1.25	1.31
0.05	1.0–1.1	1.15	1.05	1.14

Figure 6.17 shows the curves ξ (*DLF*) for different pulse durations. Additionally, a postbuckling static equilibrium path for the ideal flat plate is depicted. The presented curves (excluding the postbuckling equilibrium path) were calculated with the assumed amplitude of initial imperfections $\xi^* = 0.01$. Similarly as in the case of the curves presented in Fig. 6.16a, only a few curves in Fig. 6.17a intersect the postbuckling equilibrium path—it takes place for the curves calculated with a pulse duration less than 1.25*T* (where *T* is a period of the natural frequency of vibrations). Analysing the results presented in Table 6.11, one can see that an extension of the pulse duration leads to $DLF_{\text{cr}} = 1$, i.e., to the same results as for the static buckling load.

As has been stated previously, on the basis of the results summarized in Tables 6.8 and 6.9, it can be said that the critical dynamic load factor DLF_{cr} resulting from the proposed criterion can be found between the results obtained with application of the Volmir criterion and the Budiansky-Hutchinson criterion. The inflection point method gives similar results as for the Budiansky-Hutchinson criterion. This fact is also confirmed by the majority of the results shown in Figs. 6.16 and 6.17 and Tables 6.10 and 6.11. An advantage of the inflection point method over the Budiansky-Hutchinson criterion is a possibility to determine a specific value, and not a range of *DLF*'s in which DLF_{cr} can be found.

An exact analysis of the course of the curves presented in Figs. 6.16b and 6.17b shows that all the curves are "based" on two straight lines, one descending (for $DLF < 1$) and the other growing (for $DLF > 1$). Each of the curves r_{max} (*DLF*) has also a non-linear range, which begins after "leaving" the aforementioned straight lines. Comparing the curves and the obtained critical values DLF_{cr} for the pulse duration $T_{\text{p}} < 0.75\,T$, it can be easily seen that all these curves intersect the line for $r_{\text{max}} = 1$ in the point with the same *DLF* value. Thus, the criterion proposed by the author has a limit of applicability, because of which it cannot give the critical value of *DLF* properly for pulse durations less than 0.75 *T* in this case. Generally, it can be said that the criterion does not provide the correct value of DLF_{cr} if the curve r_{max} (*DLF*) intersects the line $r_{\text{max}} = 1$ in its linear part (for example, the curve indicated as '0.001' in Fig. 6.16b or '0.5T' in Fig. 6.17b).

Table 6.11 Influence of the pulse duration on DLF_{cr}

Pulse duration T_p	Critical value of the dynamic load factor DLF_{cr}			
	Budiansky-Hutchinson	Volmir $\xi = 1$	Author's criterion $r_{max} = 1$	Inflection point method *P-w*
0.5T	3.3–3.5	3.1	2.00	4.26
0.65T	2.3–2.5	2.24	2.00	n/a
0.7T	2.3–2.5	2.06	2.00	n/a
0.75T	2.1–2.25	1.9	1.96	n/a
1T	1.45–1.5	1.46	1.45	1.52
1.25T	1.05–1.2	1.29	1.24	n/a
1.5T	1.05–1.2	1.24	1.17	n/a

Additionally to the orthotropic plates with constant properties, some plates with material properties varying along the width of the plate were also considered. As has been presented in previous author's papers [6, 7] on static buckling and the postbuckling behaviour of structures with widthwise variable material properties, a suitable fibre distribution can lead to an increasing or decreasing buckling load.

Therefore, it was decided to repeat the buckling investigations for plates with widthwise variable material properties but now subjected to the pulse load. An impact of changes in the amplitude *A* describing the fibre volume fraction *f* (6.3) along the width of the plate on the course of the curves ξ (*DLF*) and on the critical value of the dynamic load factor DLF_{cr} was investigated. A dynamic response of plates with variable material properties subjected to rectangular pulse with a period of duration equal to a period of fundamental vibrations is shown in Figs. 6.18 and 6.19. The analysed plates are simply supported on all edges (results in Fig. 6.18) and clamped on longitudinal edges (results in Fig. 6.19). The amplitude of initial imperfection $\xi^* = 0.01$ was assumed. The results were obtained with the analytical-numerical method.

Fig. 6.18 Curves ξ (*DLF*) for the simply supported plate with a widthwise variable fibre volume fraction

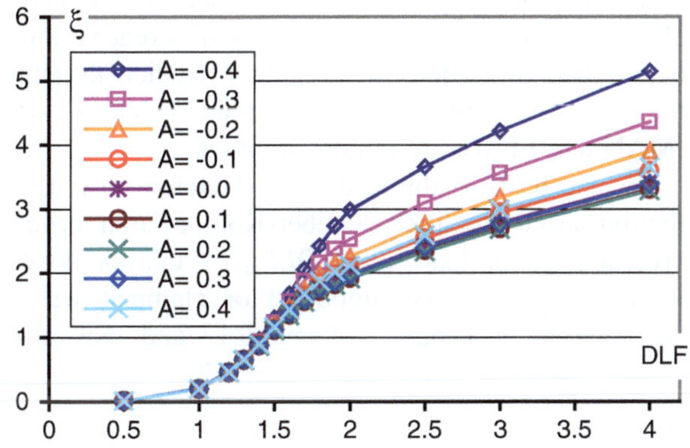

Fig. 6.19 Curves ξ (*DLF*) for the plate clamped on longitudinal edges with a widthwise variable fibre volume fraction

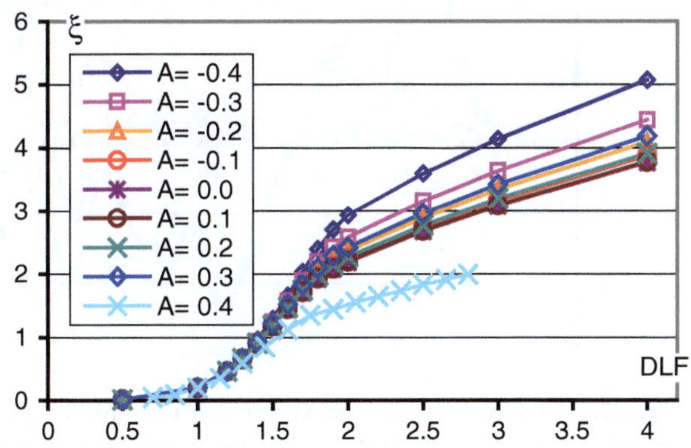

It can be noticed that the nature of the curves ξ (DLF) for all the boundary conditions taken into consideration is similar, and for small values of the dynamic load factor, i.e., $DLF < 1.5$, the course of the curves defined as the maximum deflection as a function of the dynamic load factor overlap. Only for the case in which the longitudinal plate edges are clamped (Fig. 6.19), the curve for $A = 0.4$ is different from the others. The reason of different behaviour is a different buckling mode ($m = 2$) in this case ($A = 0.4$). To check that all the calculations were correct for that case ($A = 0.4$, clamped longitudinal edges of plate), a comparative analysis was performed—the time courses of the maximum deflection were determined with the analytical-numerical method (MAN) and the finite element method (FEM) and are presented in Fig. 6.20.

The curves shown in Fig. 6.20 were determined for four values of the dynamic load factor $DLF = 1.0$, 2.0, 3.0 and 4.0. The results of both the methods employed are similar for $DLF = 1.0$, and are in accordance to the time when the deflection reached the maximum value for the remaining values of the dynamic load factor. The differences between the results obtained with both the methods became apparent for higher dynamic loads (greater *DLF*), i.e., for $DLF = 3$ or 4, and for the pulse time $t/T_p > 0.7$ (after reaching the first maximum deflection). For those *DLFs,* the buckling mode takes place, and it can be analysed with the finite element method only. Despite these differences, the critical dynamic load factor, determined by displacement criteria (Budiansky-Hutchinson or Volmir) based on the time courses of the maximum plate deflection, obtained with both the methods (MAN and FEM), are almost identical.

On the basis of the Budiansky-Hutchinson criterion, the Volmir criterion and the proposed modified Kleiber-Kotula-Saran criterion (author's criterion), the critical dynamic load factor DLF_{cr} for the composite square plate with its longitudinal edges simply supported or clamped was determined. The results are summarized in Tables 6.12 and 6.13 and they also confirm that the proposed

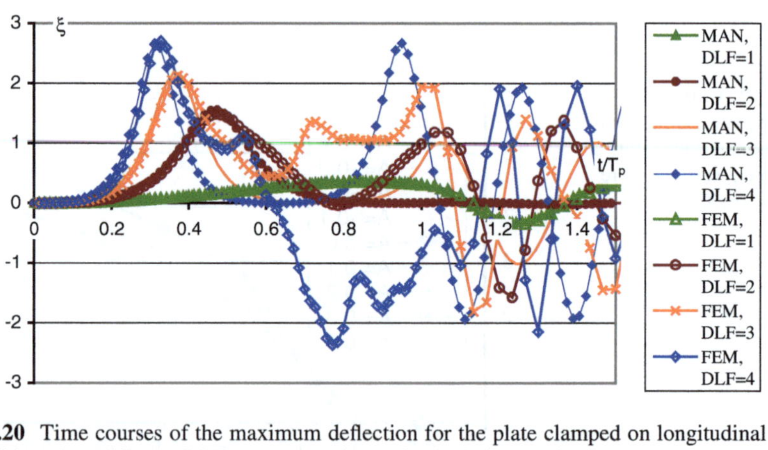

Fig. 6.20 Time courses of the maximum deflection for the plate clamped on longitudinal edges and subjected to different DLFs—a comparison of calculation methods

Table 6.12 DLF_{cr} for simply supported plates with a widthwise variable f

Amplitude of sine describing the volume fibre fraction distribution A	Critical value of the dynamic load factor DLF_{cr}		
	Budiansky-Hutchinson	Volmir $\xi = 1$	Author's criterion $r_{max} = 1$
−0.4	1.6–1.75	1.42	1.58
−0.3	1.45–1.6	1.44	1.52
−0.2	1.6–1.75	1.44	1.49
−0.1	1.45–1.6	1.46	1.47
0.0	1.3–1.45	1.45	1.44
0.1	1.3–1.45	1.47	1.44
0.2	1.45–1.6	1.47	1.43
0.3	1.3–1.45	1.47	1.45
0.4	1.45–1.6	1.45	1.47

Table 6.13 DLF_{cr} for plates clamped on longitudinal edges with a widthwise variable f

Amplitude of sine describing the volume fibre fraction distribution A	Critical value of the dynamic load factor DLF_{cr}		
	Budiansky-Hutchinson	Volmir $\xi = 1$	Author's criterion $r_{max} = 1$
−0.4	1.6–1.75	1.42	1.57
−0.3	1.45–1.6	1.43	1.53
−0.2	1.45–1.6	1.44	1.50
−0.1	1.6–1.75	1.44	1.48
0.0	1.45–1.6	1.45	1.48
0.1	1.45–1.6	1.45	1.48
0.2	1.45–1.6	1.44	1.49
0.3	1.45–1.6	1.44	1.51
0.4	1.3–1.45	1.53	1.37

criterion gives good results for plates with a relatively small amplitude of initial imperfections and subjected to pulse loading with a period of duration equal to a period of natural vibrations. The results presented in Tables 6.12 and 6.13 confirm the previously noticed rule that the critical values of the dynamic load factor DLF_{cr} determined by the proposed criterion are between the critical values obtained from the two well-known criteria (i.e., Budiansky-Hutchinson and Volmir criteria).

Analyzing exactly the results presented in Tables 6.12 and 6.13, one can also notice that the distribution of fibres along the width of the plate has no significant effect on the DLF_{cr} value (the maximum change for the same buckling mode is less than 8 %). If we exclude the case of the clamped plate with $A = 0.4$ due to a different buckling mode, a certain regularity in the value of DLF_{cr} can be observed—the critical value of DLF for the simply supported plate obtained from the Volmir criterion increases with increasing values of A from $A = -0.4$ to $A = 0.3$, and for the plate clamped along its longitudinal edges, the critical value of DLF_{cr} rises for the amplitude A approaching zero ($A = 0$ plate made of a material of constant properties). The DLF_{cr} value determined with author's criterion as well as the Budiansky-Hutchinson criterion for simply supported and clamped plates behave conversely—the critical value of *DLF* decreases for the plate with a more uniform distribution of fibres. Strengthening the middle part of the plate ($A = -0.4$) is followed by an increase in a value of the critical dynamic load factor. This is a reverse phenomenon than in the static load case (Fig. 6.4)—the largest amount of the buckling load is for the plate with $A = 0.4$ (reinforced edges).

In the majority of publications dealing with dynamic buckling problems, including the present one, the amplitude of initial imperfections $\xi^* = 0.01$ has been assumed. Such assumption is made only from the numerical point of view. Assuming such a low amplitude of initial imperfections, researchers treat the analysed structures as an almost ideal flat plate. It has been confirmed above by the earlier calculations (Tables 6.3 to 6.5) that the differences between the buckling load for the plate with a low amplitude of initial imperfections P^*_{cr} and the buckling load for an ideal flat plate P_{cr} are very small and less than 1 %.

The dynamic load factor *DLF* is defined as a ratio of an amplitude of the pulse load to the critical static buckling load for ideal structures. The calculations presented below were conducted to check how the way *DLF* is estimated influences the critical amplitude of the pulse load leading to the dynamic buckling. The author proposes to introduce a dynamic load factor $DLF^* = P/P^*_{cr}$—a pulse load amplitude divided by the static buckling load for imperfect structures. As presented in [11], such an approach is very important, especially in the case when the amplitude of initial geometrical imperfections reaches a value equal or higher than $\xi^* = 0.05$.

Some exemplary calculations showing the differences in results according to the way of the *DLF* definition, were performed for the simply supported square plate made of an epoxy-glass composite with the volume fibre fraction $f = 0.5$. Two different periods of pulse duration $T_p = T$ and $T_p = 0.5T$ (where T—period of natural fundament flexural vibrations of the plate, for the given material properties and the geometry $T = 0.59$ ms) were assumed.

To show an influence of the assumed amplitude of initial geometrical imperfections and the way of the *DLF* definition on the critical value of the dynamic load factor DLF_{cr}, a dynamic response analysis was performed with the analytical-numerical method. In Tables 6.14 and 6.15, the critical dynamic load factors DLF_{cr} (determined in the conventional way) and DLF^*_{cr} (determined from the $DLF^*(w/h)$ relations, where the amplitude of pulse loading is divided by the buckling load for the plate with initial imperfections), calculated according to the Budiansky-Hutchinson criterion, the Volmir criterion and the inflection point method, are presented.

Analysing the results presented in Tables 6.14 and 6.15, one can say that for low values of the imperfection amplitude (in a range of hundredth parts of the plate thickness), the differences between the critical value of the dynamic load factor known as DLF_{cr} (a ratio of the critical pulse load amplitude to the static

Table 6.14 DLF_{cr} and DLF^*_{cr} for different amplitudes of initial imperfections and $T_p = T$

Assumed criterion: initial imperfection amplitude ξ^*	Volmir criterion $\xi_{cr} = 1$ DLF_{cr}	Budiansky-Hutchinson criterion DLF_{cr}	Inflection point method P-w DLF_{cr}	Volmir criterion $\xi_{cr} = 1$ DLF^*_{cr}	Budiansky-Hutchinson criterion DLF^*_{cr}	Inflection point method P-w DLF^*_{cr}
0.01	1.49	1.4–1.6	1.52	1.49	1.4–1.6	1.53
0.02	1.31	1.2–1.3	1.31	1.32	1.2–1.3	1.32
0.05	1.17	0.8–0.9	1.14	1.21	0.9–1.1	1.18
0.1	1.07	0.8–0.9	1.06	1.15	0.9–1.0	1.14
0.2	1.13	0.7–0.8	0.65	1.10	0.84–0.96	1.10
0.5	0.63	0.4–0.5	0.49	1.08	0.7–0.85	0.83

Table 6.15 DLF_{cr} and DLF^*_{cr} for different amplitudes of initial imperfections and $T_p = 0.5T$

Assumed criterion: initial imperfection amplitude ξ^*	Volmir criterion $\xi_{cr} = 1$ DLF_{cr}	Budiansky-Hutchinson criterion DLF_{cr}	Inflection point method P-w DLF_{cr}	Volmir criterion $\xi_{cr} = 1$ DLF^*_{cr}	Budiansky-Hutchinson criterion DLF^*_{cr}	Inflection point method P-w DLF^*_{cr}
0.01	3.07	4.4–4.6	4.26	3.08	4.4–4.6	4.27
0.02	2.47	3.4–3.6	3.57	2.49	3.4–3.6	3.61
0.05	1.89	2.0–2.8	2.04	1.86	2.5–2.9	2.56
0.1	1.40	1.6–1.8	1.81	1.51	1.7–1.9	1.95
0.2	1.01	1.0–1.2	1.17	1.21	1.2–1.4	1.41
0.5	0.62	0.4–0.6	–	1.01	0.7–1.0	–

bifurcational load) and the proposed definition of the critical value of the dynamic load factor DLF^*_{cr} (a ratio of the critical pulse load amplitude to the static buckling load for the imperfect plate) are less than 1 % and the curves are practically identical (see Figs. 6.21, 6.22 and 6.23). However, for the initial imperfection amplitude $\xi^* \geq 0.05$, the differences between DLF_{cr} and DLF^*_{cr} grow with an increasing amplitude of initial geometrical imperfections. The calculations with a new definition of the dynamic load factor DLF* were carried out once again. The results obtained with the analytical-numerical method for the plate with all types of boundary conditions (Fig. 6.2) subjected to the rectangularly shaped pulse load with the time duration $T_p = T$ or $T_p = 0.5T$ are presented in Tables 6.16 and 6.17.

To make the differences between the results obtained on the assumption of various definitions of the dynamic load factor more visible, the courses of DLF (ξ) and DLF^* (ξ) for three values of imperfection amplitudes and two pulse durations ($T_p = 1T$ and $T_p = 0.5T$) are presented in Figs. 6.21, 6.22 and 6.23. In these figures, the static postbuckling curves P/P_{cr} (for the flat plate) and P/P^*_{cr} (for the imperfect plate) are also drawn.

It can be noticed that for a relatively small imperfection amplitude $\xi^* = 0.01$ (Fig. 6.21), the curves DLF (ξ) and DLF^* (ξ) cover each other for the given pulse

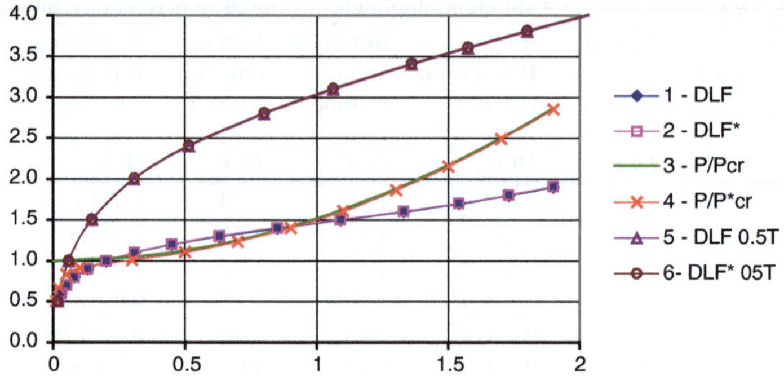

Fig. 6.21 Static and dynamic responses versus dimensionless deflection for $\xi^* = 0.01$

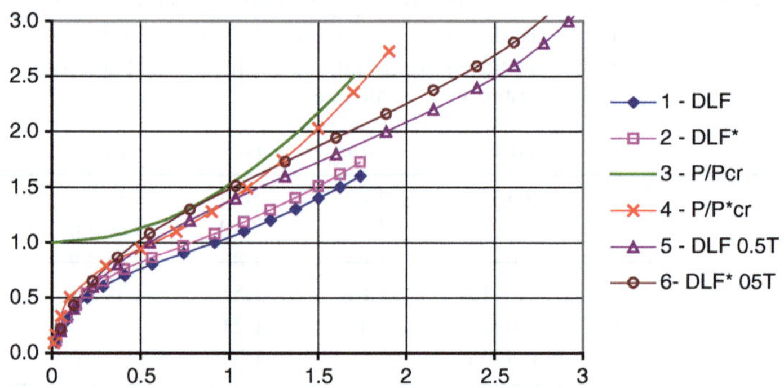

Fig. 6.22 Static and dynamic responses versus dimensionless deflection for $\xi^* = 0.1$

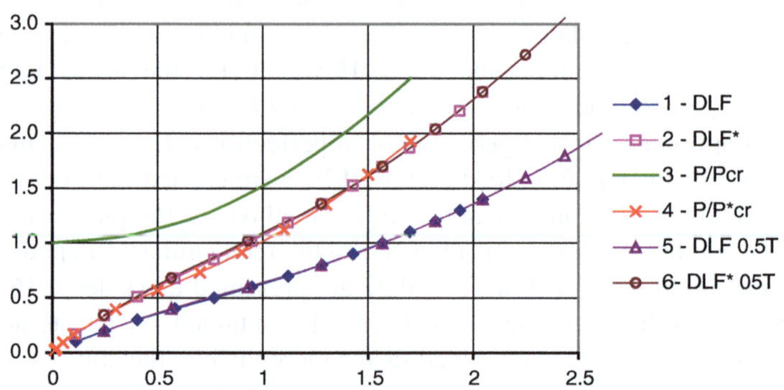

Fig. 6.23 Static and dynamic responses versus dimensionless deflection for $\xi^* = 0.5$

duration T_p. Also the static postbuckling curves overlap (excluding the initial range of deflections). In this case, the character of dynamic responses strongly depends on the assumed pulse duration—for a shorter pulse, the deflections are

Table 6.16 DLF^{*}_{cr} determined using different methods/criteria for $T_p = 1T$

Amplitude of initial imperfections ξ^{*}	Boundary conditions	Method/criteria		
		Inflection point method P-w	Volmir criterion $\xi_{cr} = 1/$ $\xi_{cr} = 0.5$	Budiansky-Hutchinson
0.01	se	1.14	1.71/1.16	1.2–1.4
0.1		0.38	1.39/0.80	0.8–1.0
0.5		0.27	0.82/0.39	0.2–0.3
0.01	ce	1.10	1.78/1.18	1.2–1.4
0.1		0.33	1.46/0.82	0.8–1.0
0.5		0.24	0.87/0.40	0.2–0.3
0.01	ss	1.45	1.40/1.06	1.4–1.6
0.1		0.68	1.04/0.70	0.8–1.0
0.5		0.43	0.63/0.35	0.4–0.5
0.01	sc	1.45	1.40/1.06	1.4–1.6
0.1		0.67	1.04/0.71	0.8–1.0
0.5		0.43	0.63/0.34	0.4–0.5
0.01	cc	1.52	1.36/1.04	1.4–1.6
0.1		0.72	0.99/0.69	0.8–1.0
0.5		0.51	0.61/0.34	0.5–0.6

Table 6.17 DLF^{*}_{cr} determined using different methods/criteria for $T_p = 0.5T$

Amplitude of initial imperfections ξ^{*}	Boundary conditions	Method/criteria		
		Inflection point method P-w	Volmir criterion $\xi_{cr} = 1/$ $\xi_{cr} = 0.5$	Budiansky-Hutchinson
0.01	se	3.92	3.32/2.36	3.8–4.2
0.1		1.10	1.52/0.93	1.2–1.4
0.5		0.29	0.82/0.39	0.3–0.4
0.01	ce	3.84	3.34/2.39	3.8–4.2
0.1		0.99	1.57/0.94	1.2–1.4
0.5		0.26	0.87/0.40	0.2–0.3
0.01	ss	4.51	3.08/2.12	4.6–5.0
0.1		1.75	1.36/0.91	1.6–1.8
0.5		0.50	0.64/0.35	0.5–0.6
0.01	sc	4.52	3.08/2.12	4.6–5.0
0.1		1.76	1.36/0.91	1.6–1.8
0.5		0.50	0.64/0.35	0.5–0.6
0.01	cc	4.56	3.01/2.07	4.6–5.0
0.1		1.87	1.34/0.90	1.8–2.0
0.5		0.59	0.62/0.35	0.5–0.6

small and the dynamic buckling load is at least three times greater (compare the results in Tables 6.14 and 6.15 or in Tables 6.16 and 6.17).

When the amplitude of imperfections grows up to the value $\xi^* = 0.1$ (Fig. 6.22), the differences between the *DLF* and DLF^* curves are clearly visible for both pulse durations and the static postbuckling curves P/P_{cr} (for the flat plate) and P/P^*_{cr} differ as well. The character of dynamic responses for the pulse durations $T_p = 1T$ and $T_p = 0.5T$ is similar but the dynamic buckling load for a shorter pulse is twice as high as for $T_p = 1T$.

For small values of the imperfection amplitude (in a range of hundredth parts of the plate thickness), the pulse load duration time strongly affects the dynamic buckling load value and the character of the dynamic response of the plate under consideration.

For a relatively large value of amplitude of initial geometrical imperfections ξ^* (the imperfection amplitude equals half of the plate thickness), the results show (Fig. 6.23) that dynamic responses of the plate do not depend on the pulse load duration—the relations *DLF* (ξ) for both the assumed periods of the pulse duration (i.e., $T_p = 1T$ and $T_p = 0.5T$) overlap and the curves DLF^* (ξ) for both the assumed pulses overlap as well. Moreover, the courses of DLF^* (ξ) are almost identical as the static postbuckling curve P/P^*_{cr}. It should be underlined that the differences between the courses of *DLF* and DLF^* are clearly visible.

One can conclude from the above mentioned observations that for considerably large values of the imperfection amplitude ($\xi^* > 0.1$), an influence of the pulse duration on the courses of *DLF* (ξ) and DLF^* (ξ) has shown to be negligible. However, the dynamic responses presented as *DLF* (ξ) and DLF^* (ξ) differ significantly. Additionally, it should be strongly underlined that the proposed relations DLF^* (ξ) calculated for two considered values of pulse load durations are almost identical with the static postbuckling curve P/P^*_{cr}, which means that the static and dynamic behaviour of the plate is practically the same for a high amplitude of initial geometrical imperfections. This fact can be only observed if the proposed definition for DLF^* is applied in the calculations.

References

1. Ari-Gur J, Simonetta SR (1997) Dynamic pulse buckling of rectangular composite plates. Compos B 28B:301–308
2. Jones RM (1975) Mechanics of composite materials. International student edition. McGraw-Hill Kogakusha, Ltd, Tokyo
3. Kelly A (ed) (1989) Concise encyclopedia of composite materials. Pergamon Press, Oxford
4. Kowal-Michalska K (ed) (2007) Statecznosc dynamiczna kompozytowych konstrukcji plytowych. WNT, Warsaw
5. Krolak M (ed) (1990) Stany zakrytyczne i nosnosc graniczna cienkosciennych dzwigarow o scianach plaskich. PWN, Warsaw – Lodz
6. Kubiak T (1998) Nieliniowa analiza statecznosci ortotropowych cienkosciennych pretow o roznych ksztaltach przekrojow poprzecznych, PhD thesis. Technical University of Lodz Press, Lodz

7. Kubiak T (2001) Postbuckling behavior of thin-walled girders with orthotropy varying widthwise. Int J Solids Struct 38(28–29):4839–4856
8. Kubiak T (2005) Dynamic buckling of thin-walled composite plates with varying widthwise material properties. Int J Solids Struct 45:5555–5567
9. Kubiak T (2005) Dynamic buckling of thin composite plates. In: Proceedings IUTAM symposium on multiscale modelling and fracture processes in composite materials, Springer, pp 123–130
10. Kubiak T (2006) Kryterium statecznosci dla konstrukcji cienkosciennych obciazonych impulsowo. In: Proceedings of stability of structure XI-th symposium, Zakopane, Poland
11. Kubiak T, Kowal-Michalska K (2012) A new approach to dynamic buckling load estimation for plate structures. In: Proceedings of Stability of Structures 13-th Symposium, Zakopane, Poland, pp 397–406
12. Petry D, Fahlbusch G (2000) Dynamic buckling of thin isotropic plates subjected to in-plane impact. Thin Wall Struct 38:267–283
13. Simitses GJ (1987) Instability of dynamically loaded structures. Applied Mech Rev 40(10):1403–1408
14. Simitses GJ (1990) Dynamic stability of suddenly loaded structures. Springer Verlag, New York
15. Simitses GJ (1996) Buckling of moderately thick laminated cylindrical shells: a review. Compos B 27B:581–587

Chapter 7
Thin-Walled Columns

The chapter deals with static buckling, postbuckling behaviour and dynamic buckling of thin-walled column with open cross-section made of isotropic and orthotropic material. The analysed columns were subjected to compressed load. The results was obtained using finite element method and analytical-numerical method.

7.1 Local Buckling of Columns with Open Cross-Section

The static and dynamic buckling of composite (epoxy glass composite with different volume fibre fractions *f*) columns with open cross-sections (Fig. 7.1) is presented in the following chapter. The assumed boundary conditions on loaded edges correspond to a simple support. The exemplary calculations were carried out for short columns with the length to web width ratio $l/b_1 = 1$ and for the following dimensions of the cross-section: $b_1/h = 50$, $b_2/h = 25$ and $b_3/h = 12.5$.

The geometrical imperfection was assumed in the shape corresponding to the static buckling mode with the amplitude $\xi^* = 0.01$.

The static buckling load and the fundamental flexural natural frequency obtained with the analytical-numerical method for columns made of a composite with different fibre fractions f are presented in Tables 7.1 and 7.2. The static postbuckling equilibrium paths for columns with a few cross-sections under consideration (Fig. 7.1), with different fibre volume fractions, are presented in Figs. 7.2 and 7.3. The postbuckling equilibrium paths are calculated for columns with ideal flat walls (Fig. 7.2a) and for columns with the initial imperfection corresponding to the buckling mode with an imperfection amplitude equal to 1/100 of the column wall thickness (Figs. 7.2b and 7.3)

The buckling load and the natural frequency for the hat column section and the lipped column section are similar—it is true only for the local buckling case. Buckling for the channel section column was caused by flanges—this is a reason why for this cross-section the buckling load and the natural frequency are smaller

T. Kubiak, *Static and Dynamic Buckling of Thin-Walled Plate Structures*,
DOI: 10.1007/978-3-319-00654-3_7,

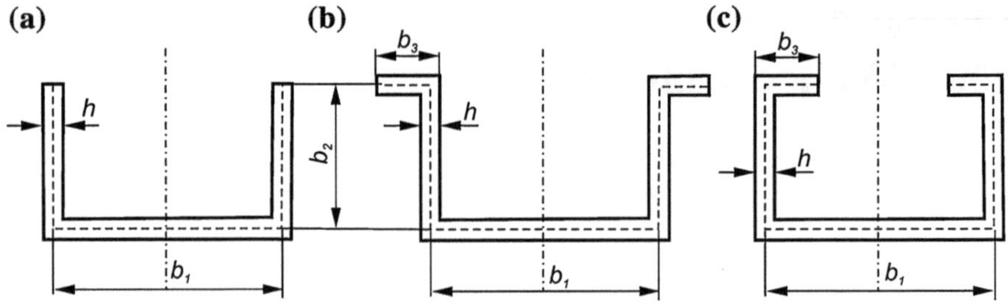

Fig. 7.1 Cross-sections of the columns under consideration

than for two other analysed cross-sections. Looking at the obtained results (Tables 7.1 and 7.2), one can see that an increase in the volume fibre fraction f leads to an increase in buckling loads as well as in natural frequencies. Postbuckling equilibrium paths for the lipped channel section and the hat channel section overlap. The postbuckling path for the channel section lies below the equilibrium paths of columns with stiffened cross-sections—it is obvious because columns with stiffened cross-sections have similar stiffness and channel section columns are more flexible.

Postbuckling equilibrium paths for columns with different volume fibre fractions f almost overlap (see Fig. 7.3). However, the curves representing the columns with $f = 0.7$ lie slightly below the others. The above observation could be explained by differences in coefficients of orthotropy, calculated as Young's modulus in the longitudinal direction divided by Young's modulus in the transverse direction, which are as follow: 3.4 for $f = 0.3$; 3.6 for $f = 0.5$ and 3.1 for $f = 0.7$. If one makes a zoom on the results presented in Fig. 7.3a (Fig. 7.4) it is well visible that the course of postbuckling curves depends on the coefficient of orthotropy.

The dimensionless deflections $\xi = w/h$ (the maximum displacement divided by the segment wall thickness) as a function of *DLF* (dynamic load factor) for different shapes of pulses are presented in Figs. 7.6, 7.7, 7.8, 7.9 and 7.10. The results presented in these figures are obtained for:

- the pulse loading duration T_p equal to the period of natural vibrations corresponding to the mode $m = 1$;
- the shape imperfection corresponding to the local buckling mode with an amplitude equal to 1/100 of the segment wall thickness ($\xi^* = 0.01$);
- different shapes of the pulse loading—they are presented in Fig. 7.5.

Table 7.1 Buckling load P_{cr} (N) for different cross-sections and volume fibre fractions

Volume fibre fraction f: cross-section	0.2	0.4	0.6	0.8
Channel section (Fig. 7.1a)	920	1526	2281	3484
Hat channel section (Fig. 7.1b)	1796	2819	4214	6753
Lipped channel section (Fig. 7.1c)	1797	2821	4217	6762

Table 7.2 Natural frequency (Hz) for different cross-sections and volume fibre fractions

Volume fibre fraction *f*:cross-section	0.2	0.4	0.6	0.8
Channel section (Fig. 7.1a)	786	940	1076	1256
Hat channel section (Fig. 7.1b)	982	1142	1308	1563
Lipped channel section (Fig. 7.1c)	983	1142	1308	1564

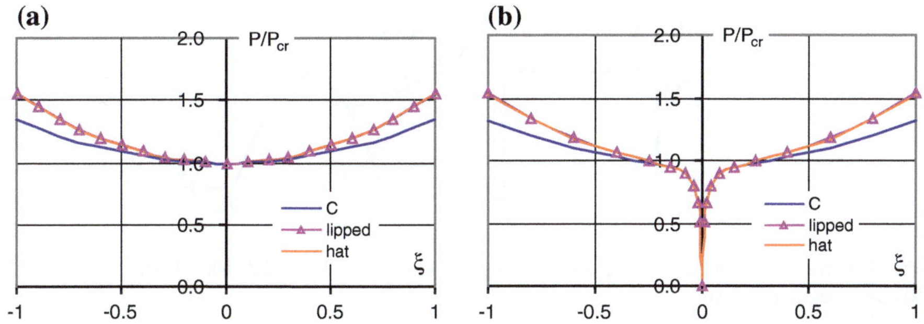

Fig. 7.2 Postbuckling equilibrium paths for the columns under consideration and the volume fibre fraction $f = 0.5$

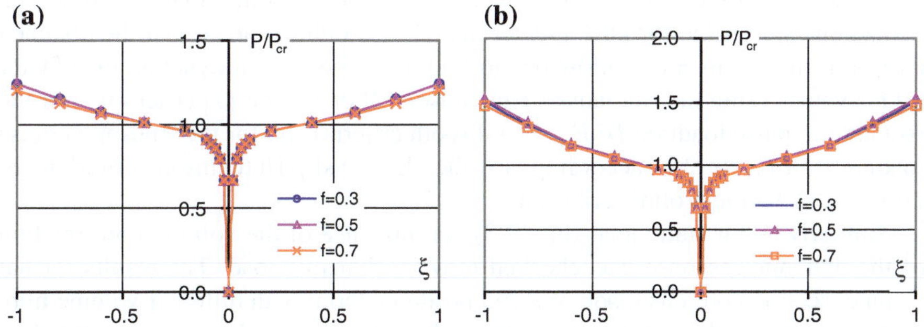

Fig. 7.3 Postbuckling equilibrium paths for the channel section (**a**) and hat channel section (**b**) columns with different volume fibre fractions

Fig. 7.4 Postbuckling equilibrium paths for hat channel section columns with different volume fibre fractions (zoom of Fig. 7.3b)

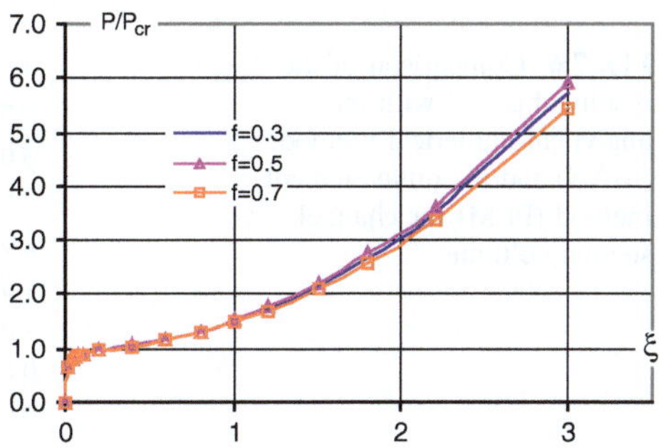

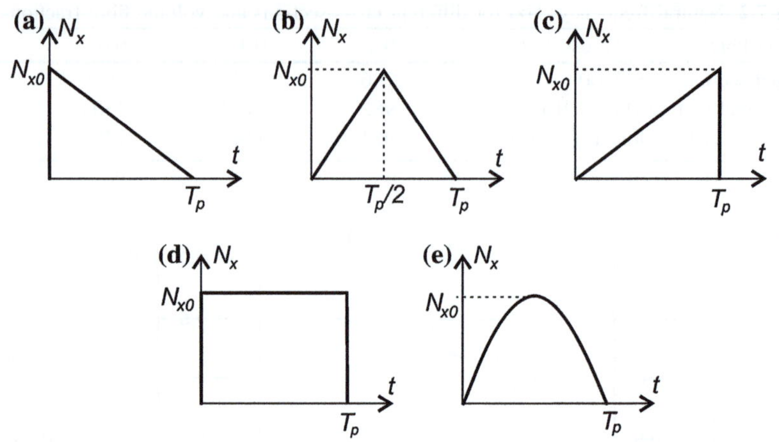

Fig. 7.5 Shapes of pulse loading: three triangular impulses T1 (**a**), T2 (**b**), T3 (**c**), rectangular R (**d**) and sinusoidal S (**e**)

The results presented in Fig. 7.6 were obtained with the analytical-numerical method (MAN) and compared with FEM computations. They are similar for both assumed shapes of pulse loading (Fig. 7.5): rectangular (R) and sinusoidal (S). Nevertheless, for the rectangularly shaped pulse loading, some small differences in deflections are visible for *DLF* greater than 2. It should be noted that the obtained curves from both methods allow one to find the same critical dynamic load factor DLF_{cr} values using the Budiansky-Hutchinson [2] or Volmir [7] criterion—for the rectangular pulse loading: $DLF_{cr} \approx 1.4$ (both criteria) and for the sinusoidal pulse loading: $DLF_{cr} \approx 2.1$ according to the Budiansky-Hutchinson criterion or $DLF_{cr} \approx 2.3$—the Volmir criterion.

Similarly as for static load (Fig. 7.3), an influence of the volume fibre fraction of the dynamic response was checked also for dynamic load. The results for hat channel section columns made of a composite material with different volume fibre fractions subjected to the rectangularly shaped pulse loading are presented in Fig. 7.7. All curves presented in Fig. 7.7a lie very close to each other, and the curves overlap for *DLF* less than 1.8. The above observation leads to a conclusion that the Budiansky-Hutchinson or Volmir criterion for all the analysed volume

Fig. 7.6 Comparison of the results obtained with the analytical-numerical method (MAN) and the finite element method (FEM) for channel section columns

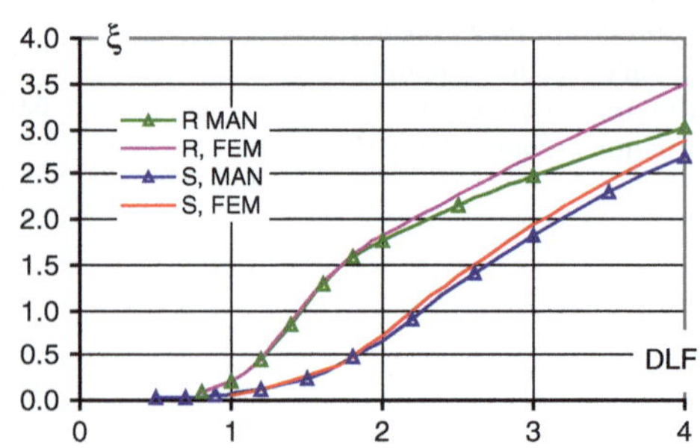

fibre fractions (Fig. 7.7a) has yielded almost the same value of the critical dynamic load factor DLF_{cr}. But it should be noted that the same DLF_{cr} does not mean the same critical amplitude of the applied pulse (compare Figs. 7.7a and 7.7b). Concluding the above, one can say that given channel section columns with different volume fibre fractions behave similarly during the dynamic response as the column under static load during its postbuckling behaviour (compare the results presented in Figs. 7.4 and 7.7a).

In Fig. 7.8, a comparison of the dynamic response of columns made of a composite material ($f = 0.5$) with different cross-sections subjected to the triangular T3 pulse is presented. The curves for the hat channel section and the lipped channel section cover each other. These results are similar to the postbuckling behaviour for the same column subjected to static load—compare the curves presented in Figs. 7.2 and 7.8.

Summing up the above-mentioned, it can be said that the relations observed in the postbuckling analysis and based on postbuckling equilibrium paths are similar to the dynamic response analysis based on the dimensionless deflection versus the dynamic load factor.

For other shapes of the pulse load (Fig. 7.9), the courses of curves are similar to that presented in Fig. 7.8 for the triangular T3 shape pulse load. In Fig. 7.9, a comparison of the dynamic response of columns made of a composite material ($f = 0.5$) with different cross-sections subjected to the triangular T1 and T3 pulse is presented.

Dynamic responses for lipped channel section columns for different pulse loads are presented in Fig. 7.10. The curves denoted by S = R were obtained for the sinusoidal pulse loading with the same area as the rectangular pulse loading (for the same pulse duration, the amplitude was higher for the sinusoidal pulse). The

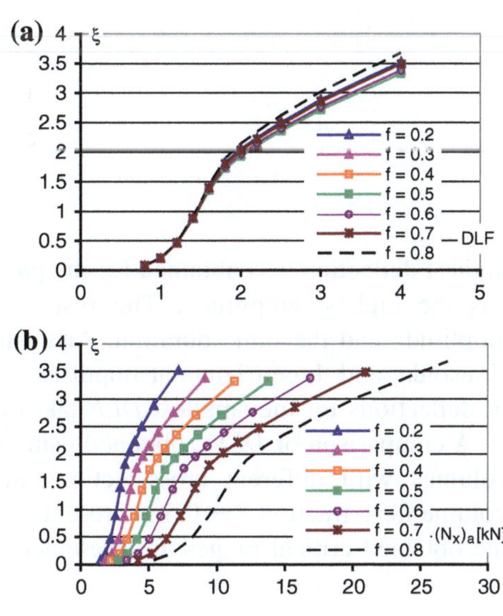

Fig. 7.7 Dimensionless deflection versus *DLF* (**a**) and the dimensionless deflection versus the pulse amplitude $(N_x)_a$ (**b**) for the hat channel section column

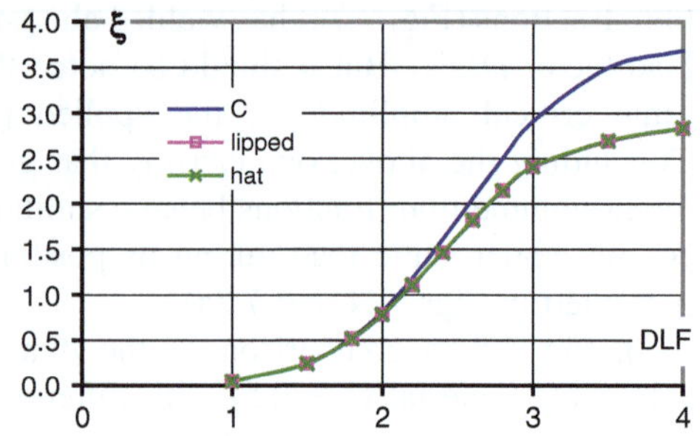

Fig. 7.8 Columns subjected to the T3 pulse loading—a comparison of cross-sections

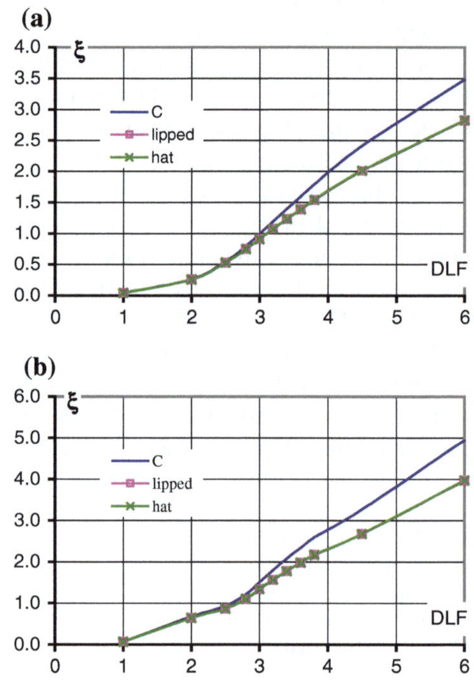

Fig. 7.9 Columns subjected to the T1 (**a**) and T2 (**b**) pulse loading—comparison of cross-sections

highest deflection was obtained for the pulse denoted by S = R because that pulse has the highest amplitude. The rest of the compared pulses have the same amplitude and the same duration. Analysing the curves ξ (*DLF*) for rectangular, sinusoidal and three triangular impulses, it can be seen that the highest increment of deflections for the smallest *DLF* takes place for the rectangular pulse loading.

A comparison of DLF_{cr} obtained using the Budiansky-Hutchinson criterion for columns with different cross-sections made of composite materials with the volume fibre fraction $f = 0.5$ is presented in Table 7.3. Only average values from the obtained critical ranges are presented. The dynamic load factors for different

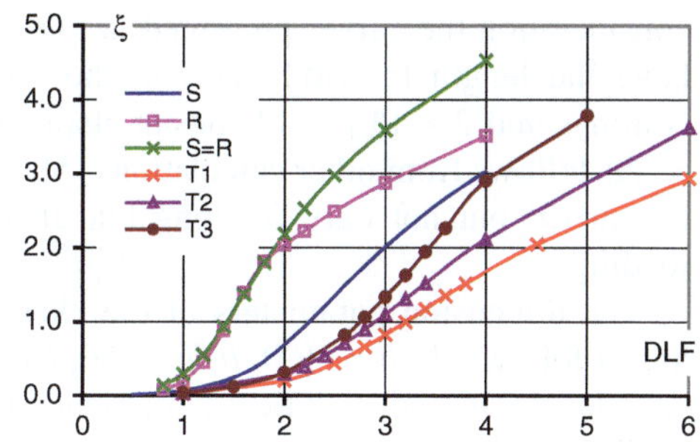

Fig. 7.10 Comparison of different shapes of the pulse loading for lipped channel section composite columns with the fibre volume fraction $f = 0.7$

cross-sections are in the same relation as buckling loads (Table 7.1)—the same or similar DLF_{cr} for cross-sections with stiffeners (hat channel section and lipped channel section).

7.2 Interactive Buckling of Columns with Open Cross-Sections

In this section, the interactive dynamic buckling of short and medium length columns is presented. All the results presented in this section were obtained for structures made of steel with material properties data given in Table 6.1.

First, short columns with channel sections (Fig. 7.1a) and hat channel sections (Fig. 7.1b) subjected to the rectangular pulse loading (Fig. 7.5d) are considered. A short column is defined as a column for which the primary buckling mode $m = 1$ is the local mode (longitudinal edges of the column are straight) and the shape ratio (the length of walls to their width) for the widest wall is greater than 2 [6]. A range of length for such defined columns depends on their cross-section. For the case under analysis, the length of "short columns" can be very easily found by means of the curves presented in Fig. 7.11, where the curves denoted by "m = 1" were obtained for the buckling mode corresponding to one halfwave of the sinusoid, and the curves denoted by "m > 1" were obtained for the local buckling mode (number of halfwaves of the sinusoid is greater than one). Taking into

Table 7.3 DLF_{cr} value according to the Budiansky-Hutchinson criterion for different shapes of applied pulses

Analysed cross-section: type of pulse	Channel section	Hat channel section	Lipped channel section
S	2.0	2.1	2.1
R	1.6	1.6	1.6
S = R	1.4	1.4	1.4
T1	3.1	3.1	3.1
T2	2.3	2.5	2.5
T3	2.4	2.5	2.5

consideration the curves presented in Fig. 7.7, it can be said that “short columns” have the length to width ratio in the range $2 < l/b_1 < 10$ for channel section columns and $2 < l/b_1 < 14$ for hat channel section columns.

As follows from other calculations [5], the multimodal buckling analysis should be carried out not only for long but also for short columns subjected to pulse loading.

The dimension parameters of the exemplary column taken into consideration are as follows: $b_1/h = 100$, $b_2/h = 50$, $b_3/h = 25$ and $l/b_1 = 4$. The problem was calculated with the analytical-numerical method and the finite element method [8].

To explain why an interaction of different buckling modes should be taken into account in the dynamic response analysis of short channel section columns, the results obtained with the finite element method are presented first.

It was observed that during the analysis of deformation of the channel section column, the edge between the web and the flange was also deformed—the edges do not remain straight as it is in the case of the local buckling mode (see Fig. 7.12).

The results presented in Fig. 7.12 were obtained on the following assumptions:

- the shape of initial imperfections corresponded to the local buckling mode $m = 3$;
- the amplitude of initial imperfections was equal to 1/100 of the column wall thickness;
- the pulse duration was equal to a period of natural vibrations with the modal mode corresponding to the local buckling mode, i.e., $T_p = T_3 = 1.6$ ms.

During the pulse load, the edge deflection (curve 2—Fig. 7.12) is greater than the deflection of the middle point of the web (curve 1—Fig. 7.12). The above observation means that not only the local but also global buckling mode appeared in the case of pulse loading. Taking the above into account, a graph representing dimensionless deflection as a function of the dynamic load factor has been prepared employing the deflection calculated in points 1 and 2 (Fig. 7.12)—the results

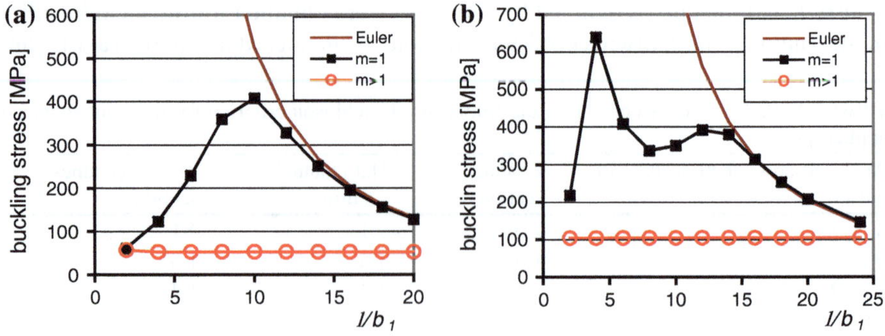

Fig. 7.11 Buckling stress versus the length parameter for the channel section (**a**) and hat channel section (**b**) columns

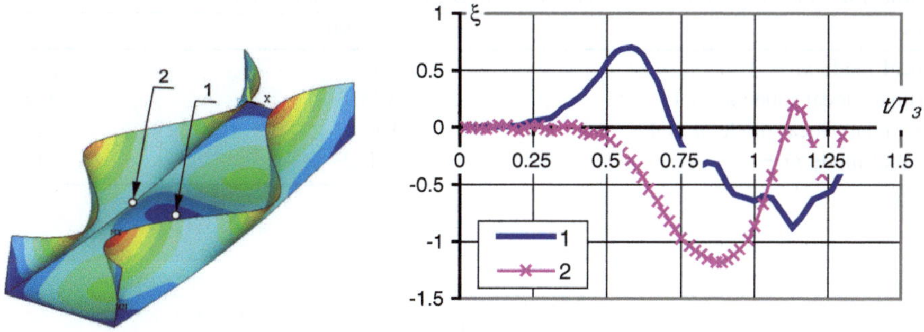

Fig. 7.12 Deflection versus time for the channel section column subjected to the rectangular pulse loading with $DLF = 1.6$

are presented in Fig. 7.13. The course of the curve denoted by 1 has the character corresponding to the local dynamic buckling mode and the character of curve 2 has the course corresponding to the global buckling mode. The channel section column behaviour means that an interaction between different modes appeared and it should be taken into account in the analytical-numerical method.

The results obtained from FEM give the information on the interactive dynamic buckling of the short channel section column. In order to find a possible mode included in the interaction, a linear buckling analysis and a modal analysis were performed. The obtained results are summarized in Table 7.4 and presented in Fig. 7.14.

Taking into account all four buckling modes (presented in Table 7.4 and in Fig. 7.14) in the analytical-numerical method allows one to obtain results similar to those obtained with FEM. A comparison between the results obtained with the analytical-numerical method and the finite element method on the plots presenting dimensionless deflections versus the dynamic load factor for the channel section column are shown in Fig. 7.15. Some differences appear because the model has only a few degrees of freedom in the employed analytical-numerical method. The curves presented in Fig. 7.15 are different but the critical dynamic load factor obtained from these curves with the Budiansky-Hutchinson criterion is similar.

Fig. 7.13 Dimensionless deflection at point 1 and 2 of the channel section column versus the dynamic load factor

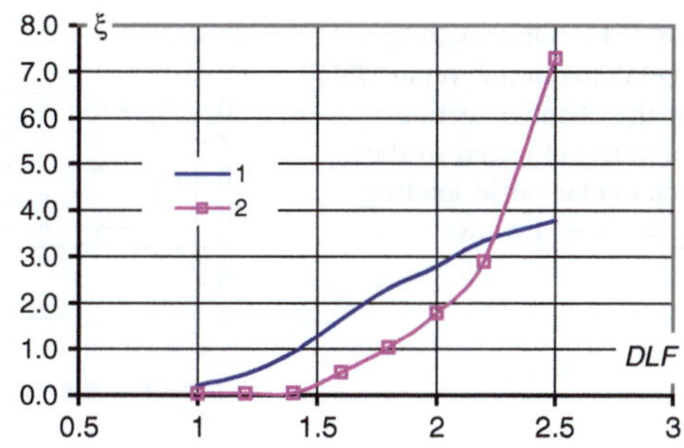

Table 7.4 Buckling stress and natural vibrations for the channel column

Mode	σ_{cr} (MPa)	n (Hz)
Local mode $m = 3$	53	614
Primary local mode $m = 1$	123	312
Secondary local mode $m = 1'$	972	880
Global mode $m = 1''$	5,122	2,001

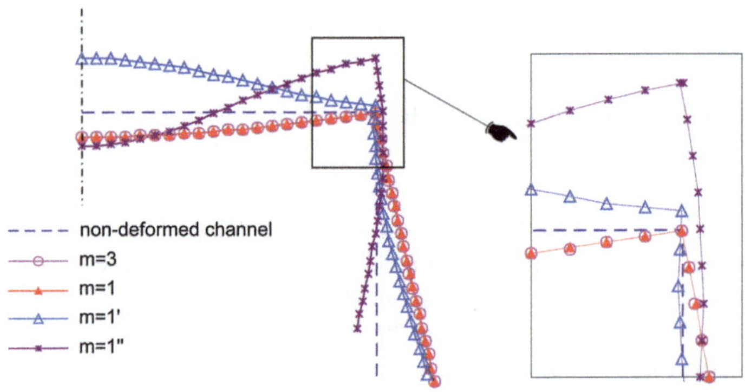

Fig. 7.14 Buckling modes for the channel section column

Fig. 7.15 Dimensionless edge deflection ξ versus *DLF* for the channel section columns subjected to the rectangular pulse loading $T_p = T_3 = 1.6$ ms

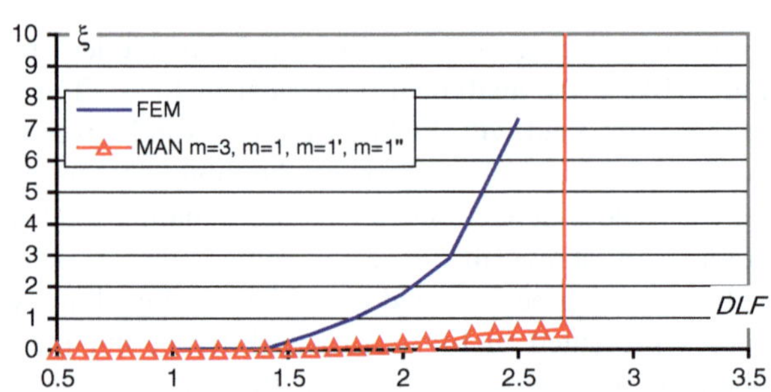

Fig. 7.16 Dimensionless edge deflection ξ versus *DLF* for the channel section columns subjected to the rectangular pulse loading $T_p = T_3 = 1.6$ ms

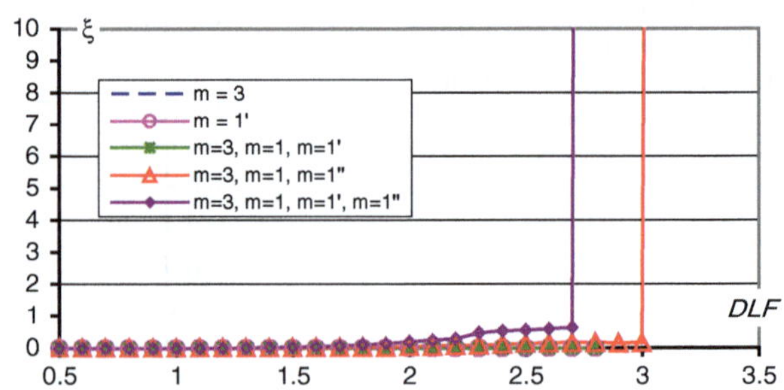

From the analytical-numerical method, $DLF_{cr} = 2.7$ and from the FEM, $DLF_{cr} = 2.6$.

If not all buckling modes presented in Table 7.4 are not taken into account in the analytical-numerical some wrong results can be obtained—the character of the global buckling mode did not appear (see Fig. 7.16)

As presented in Fig. 7.16, a channel section column deflection with the character corresponding to the global buckling mode could be obtained in the analytical-numerical calculation if the global mode "$m = 1$" (Fig. 7.14) is taken into consideration together with local modes.

It should be noted that the finite element method gives results even in the case when only one buckling mode as the initial imperfection (for example, the local buckling mode $m = 3$) is taken into account. This gives an advantage of the finite element method, however, the use of analytical-numerical methods allows one to specify which modes should be taken into consideration (dynamic interactive buckling) in the calculation to obtain satisfactory results (similar for both the methods).

Thin-walled plated structures subjected to static load can work after a local loss of stability. Because this is valid also for the structures subjected to the pulse load for which the first local buckling occurs, it is necessary to analyze the behaviour of structures at dynamic loads larger than the critical amplitude of the pulse causing a local loss of stability. As has been demonstrated in the example shown above, the global dynamic buckling mode appears for the dynamic load factor *DLF* about twice as high as DLF_{cr} for the local dynamic buckling mode (for static loads, this relationship in the present case is about 100—see Table 7.4).

It should be also noted that:

- the *DLF* coefficient is referred to the lowest buckling load values (the buckling load for the local mode $m = 3$);
- the critical stress for the global mode ($m = 1'$ and $m = 1''$) taken into account in the multimodal analysis is approximately 18 and 100 times greater than the minimum local buckling mode.

This observation demonstrates a need to select the appropriate buckling mode of the theoretical analysis of the interactive buckling (multimodal), following not only the knowledge of the interactive buckling of structures under static loads. As demonstrated by the results presented here, not only natural frequencies with the mode corresponding to the buckling mode but also the ratios of natural frequencies should be taken into account in the interactive buckling analysis. In the present case, the ratio of periods of vibrations with the mode corresponding to the following buckling modes $m = 3$, $m = 1$, $m = 1'$ and $m = 1''$ can be written as 2:1:2.8:6.4.

Let us examine the behaviour of the considered channel section steel column subjected to a rectangular pulse load with twice as long duration than previously, i.e., for a period of natural vibrations with a mode corresponding to the mode $m = 1 - T_p = T_1 = 3.2$ ms. The graphs presenting the dimensionless maximum deflection of the column edge (point "2" in Fig. 7.12) and the column web (point "1" in Fig. 7.12) as a function of the dynamic load factor are shown in Figs. 7.17 and 7.18. *DLF* is defined in relation to the lowest static local buckling load.

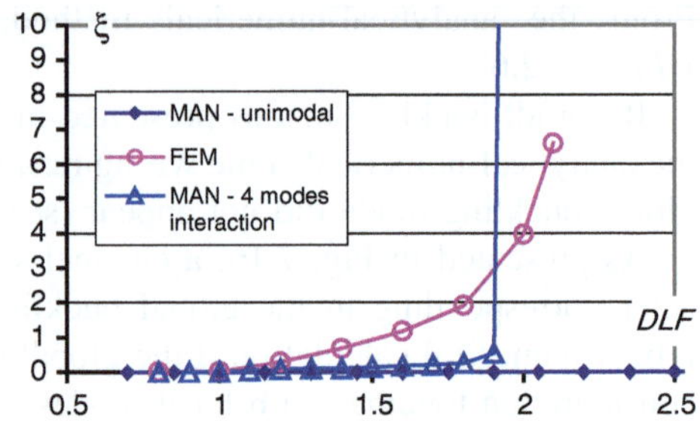

Fig. 7.17 Edge of the column dimensionless deflection ξ as a function of *DLF* for a pulse load with the period of duration $T_p = T_1 = 3.2$ ms

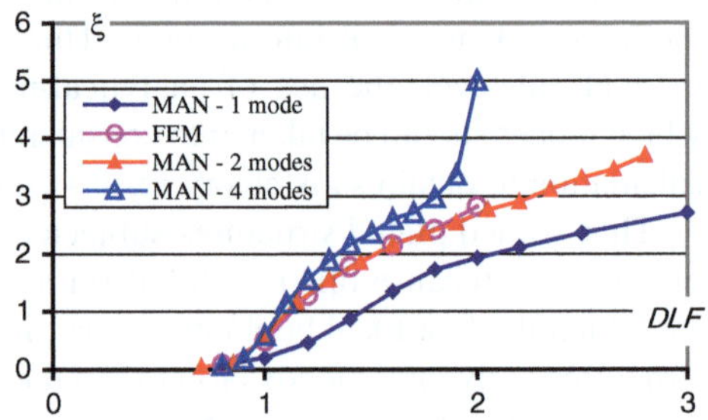

Fig. 7.18 Middle point of the column web dimensionless deflection ξ as a function of DLF for a pulse load with the period of duration $T_p = T_1 = 3.2$ ms

Also in this case, when the global buckling mode is analysed (Fig. 7.17), the rising character of the dimensionless deflection resulting from the two employed methods is different, but the critical values of dynamic load factors, obtained after applying the Budiansky-Hutchinson criterion, are similar—$DLF_{cr} = 2.1$ obtained with the FEM and $DLF_{cr} = 1.9$ from the analytical and numerical method.

Analysing the results shown in Fig. 7.18, it can be concluded that the designation of the critical load factor is sufficient to take only two modes into the interactive analysis, for which the critical dynamic load factor DLF_{cr} according to the Budiansky-Hutchinson criterion, is about 1.1 and is consistent with the results obtained from the analytical-numerical interactive analysis with four modes taken into account and with the results obtained with FEM.

To show that not all short columns with open cross-sections require an interactive buckling analysis, some exemplary FEM calculations of the hat channel section column (Fig. 7.1.b) with the cross-section dimension $b_1/h = 100$, $b_2/h = 50$, $b_3/h = 25$ and the length parameter equal to $l/b_1 = 4$ are presented below.

The calculations were performed on the following assumptions:

- the shape of initial imperfections corresponded to the local buckling mode $m = 5$;
- the amplitude of initial imperfections was equal to 1/100 of the column wall thickness;
- the pulse duration was equal to a period of natural vibrations with the modal mode corresponding to the local buckling mode, i.e., $T_p = T_5 = 1.6$ ms.

The dimensionless deflection ξ (the deflection divided by the wall thickness for the characteristic point of the middle column cross-section) as a function of dimensionless time (time divided by pulse duration) for the hat channel section column is presented in Fig. 7.19. The characteristic points are located in the middle cross-section of the column and they are (see Fig. 7.19) in the middle of the web (point 1), on the edge between the web and the flange (point 2) and on the free edge of the flange stiffener (point 3).

Comparing the results presented in Fig. 7.19, it can be seen easily that only local buckling occurs—the deflections in points 2 and 3 are negligibly small in comparison with the deflection in point 1. The edge (point 2) deflection higher than two thicknesses of the column wall appears for *DLF* greater than 6.5 (see curve 2 in Fig. 7.21). An example showing that a deflection of the edge is similar to the web deflection even for $DLF = 6$ is presented in Fig. 7.20.

Dimensionless deflections ξ as a function of the dynamic load factor DLF of the hat channel section column established for characteristic points (see Fig. 7.19) are presented in Fig. 7.21. These curves allow one to determine the dynamic buckling load (the critical dynamic load factor—DLF_{cr}) with a particular dynamic buckling criterion. Due to the local buckling character of column deflections, curve 2 was taken into account for determining DLF_{cr}. According to the Volmir criterion, the critical value of the dynamic load factor equals 1.5, using the Budiansky-Hutchinson criterion, the DLF_{cr} for local buckling is in the range from 1.5 to 1.6.

Taking into consideration curve 2 in Fig. 7.21, it can be seen that for $DLF = 5$, deflections are smaller than for $DLF = 4$. It means that the buckling mode changed at an increasing *DLF* (see Fig. 7.22. It follows that the Ari-Gur [1] criterion can be used. DLF_{cr} determined with the Ari-Gur criterion is equal to 4.

Let us consider a slightly longer column, i.e., a column with intermediate length. This type of column could be defined as a column whose buckling mode with the number of halfwaves equal to 1 ($m = 1$) corresponds to the global mode and differences between the local and global buckling load are less than 4 ($P_{cr}^{local}/P_{cr}^{global} < 4$), with the lowest buckling load corresponding to the local mode.

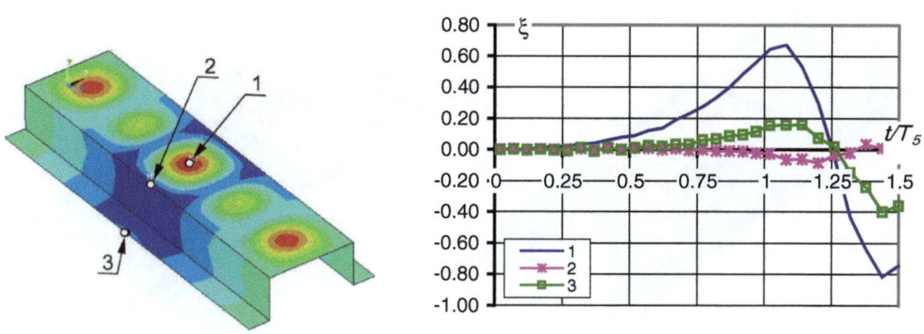

Fig. 7.19 Deflection versus time for the hat channel section column subjected to the rectangular pulse loading at $DLF = 1.6$

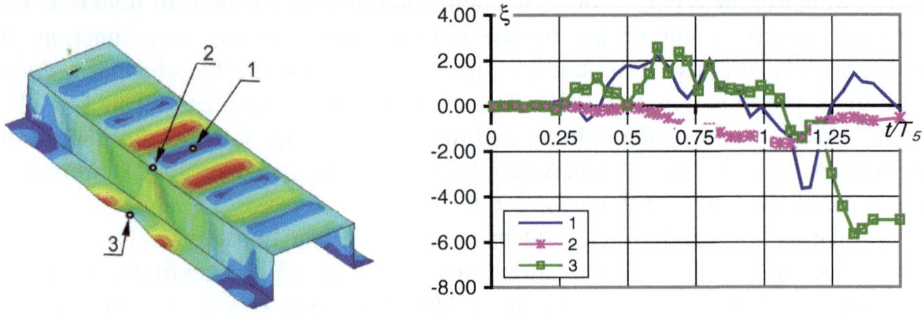

Fig. 7.20 Deflection versus time for the hat channel section column subjected to the pulse loading at $DLF = 6$

Fig. 7.21 Dimensionless deflection versus *DLF* for the channel section column subjected to the rectangular pulse loading

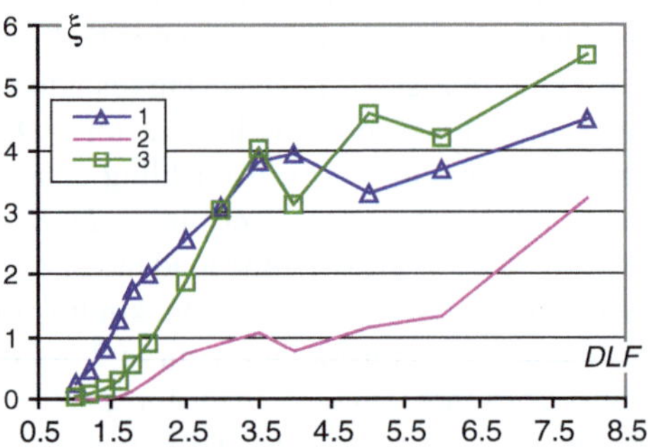

For such a column, an interactive buckling analysis is necessary and more than two buckling modes should be considered in numerical calculations [3, 4].

Exemplary results of the interactive dynamic buckling analysis performed with FEM and the analytical-numerical method (MAN) for a simply supported steel channel section (Fig. 7.1a) column with the following dimension parameters $b_1/h = 50$; $b_2/h = 25$, $l/b_1 = 8$ are presented below. The considered column was

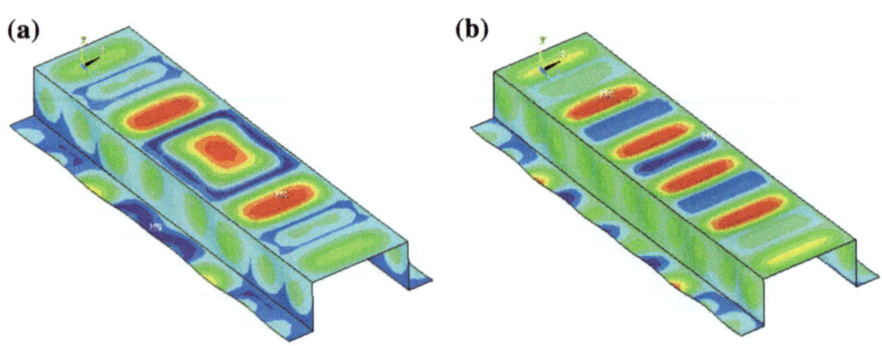

Fig. 7.22 Deflection of the column with an omega cross-section for different dynamic load factors $DLF = 4$ (**a**) and $DLF = 5$ (**b**)

subjected to uniform compression in the form of the rectangularly shaped pulse loading.

The lowest static buckling load corresponding to the local mode ($m = 6$) is equal to $P_{cr6} = 21.0$ kN and the buckling load corresponding to the global flexural mode ($m = 1$) is equal to $P_{cr1} = 71.3$ kN. The above value is useful for calculating the dynamic load factor which in this case is calculated as an amplitude of the pulse load divided by the local buckling load $P_{cr6.}$

The modal analysis allows one to calculate natural frequencies for the mode corresponding to the buckling mode, and they are $n_1 = 385$ Hz and $n_6 = 1258$ Hz for local ($m = 6$) and global ($m = 1$) modes, correspondingly. The above results allow one to determine a period of the assumed pulse duration which could be $T_1 = 1/n_1 = 2.59$ ms or $T_6 = 1/n_6 = 0.79$ ms.

In the case when the pulse load duration is more than 1.5–2 times longer than the free vibration period corresponding to the given mode, then such a load can be treated as a quasi-static load. Because for the considered column $T_1/T_6 = 3.25$, then under the pulse duration assumed as $T_p = T_1$, the applied load should be treated as the dynamic one for the global buckling mode, and as the quasi-static one for the lowest local buckling mode ($m = 6$).

In Fig. 7.23, results of the uncoupled buckling analysis performed with the analytical-numerical method (MAN) are presented as plots of maximum dynamic dimensionless deflections as a function of the dynamic load factor. It refers to two buckling modes under consideration, that is to say, for $m = 1$ and $m = 6$. The period of the pulse duration was set as $T_p = T_6 = 0.79$ ms.

According to the dynamic stability criterion, proposed by Budiansky and Hutchinson for uncoupled dynamic buckling, from the graphs in Fig. 7.23, the critical value of the dynamic load factor can be established as $DLF_{cr} = 1.2$ for the local buckling mode and $DLF_{cr} = 4.8$ for the global buckling mode.

Next, the dynamic analysis of coupled buckling of three buckling modes (the global one and the two local ones—the primary and secondary local modes for $m = 6$) was conducted using the analytical-numerical method. The dynamic load factor was defined as a multiplication of the local buckling load for the mode $m = 6$. The initial imperfection amplitudes of interactive modes in the analytical-numerical method are assumed as follows: $\xi_1^* = 0.05$ for the global mode, $\xi_6^* = 0.05$ for the primary local mode, and $\xi_{6'}^* = 0$ for the secondary modal mode.

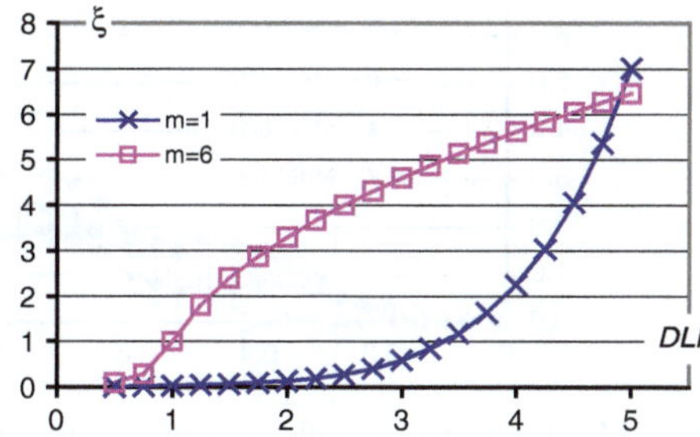

Fig. 7.23 Maximum dynamic dimensionless deflection versus the dynamic load factor for the uncoupled dynamic buckling analysis

During FEM computations, the initial imperfection was assumed only with the shape corresponding to the local buckling mode with the amplitude $\xi_6^* = 0.05$. To be sure that the global mode during dynamic load decides about the column deflection, the pulse duration was set as a period of natural frequencies with the mode corresponding to the global buckling load $T_p = T_1 = 2.59$ ms.

A comparison of the results obtained with FEM and the analytical-numerical method (MAN) is presented in Figs. 7.24 and 7.25. These graphs show dimensionless dynamic deflections in the point lying on the web along the symmetry axis of the column with dimensionless coordinates equal to $x/l = 1/12$, 5/12 and 1/2. The dynamic response (the column dimensionless deflection ξ) of the impacted channel section column subjected to a rectangularly shaped pulse with the dynamic load factor $DLF = 1.2$ is presented in Fig. 7.24 and with $DLF = 1.5$—in Fig. 7.25. The results obtained with the analytical-numerical method, which is based on the plate theory, are valid only when the deflection is less than 10 times the thickness of the plate. For that reason, the graphs presented in Figs. 7.24 and 7.25 are scaled for ξ from 0 to 10.

From the comparison of the curves in Figs. 7.23 and 7.24, a rapid increase in deflections in the case of the interactive dynamic buckling analysis is visible, whereas from the analogous comparison of Figs. 7.23 and 7.25, a dramatic increase in deflections leading to failure of the structure (Fig. 7.26) can be observed.

Figure 7.26 shows a dynamic displacement of coupled buckling of the channel section column subjected to the pulse compression load for four different moments of time defined as t/T_1 and for the dynamic load factor $DLF = 1.5$.

As can be seen in Fig. 7.26c for the dimensionless time value $t/T_1 = 1$, the maximum dimensionless deflection reaches the value $\xi = 50$, and for $t/T_1 = 1.5$ (Fig. 7.26d) $\xi = 135$.

Comparing the results obtained with the finite element method (FEM) and the analytical-numerical method (MAN), it can be said that a good agreement is achieved. This suggests that the analytical-numerical method MAN presented in Chap. 3 allows one to analyse thin-walled plate structures not only for uncoupled

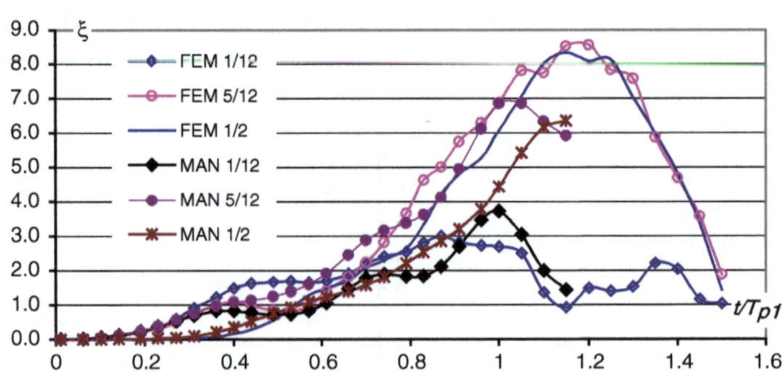

Fig. 7.24 Dimensionless deflection versus dimensionless time for $DLF = 1.2$

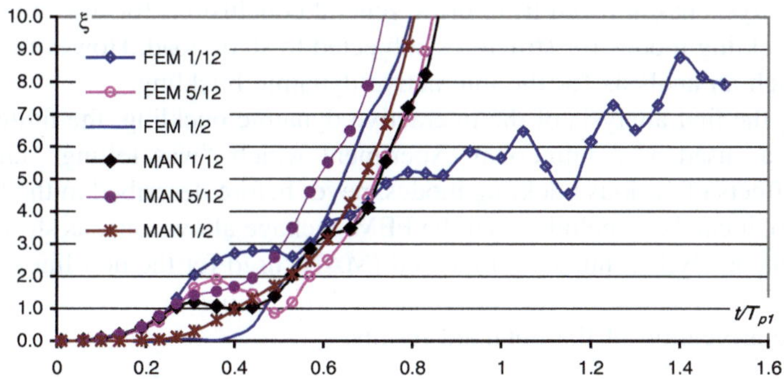

Fig. 7.25 Dimensionless deflection versus dimensionless time for $DLF = 1.5$

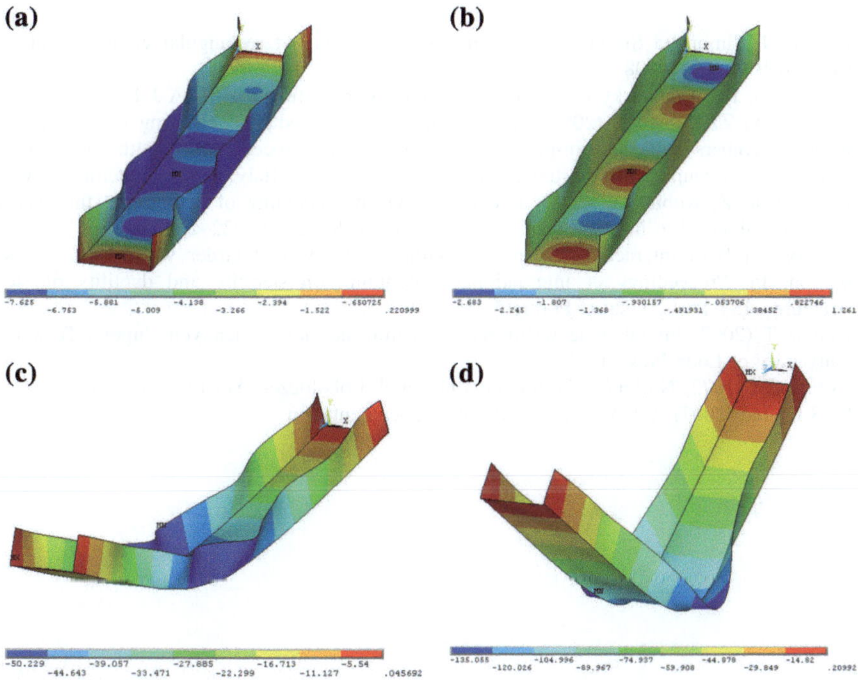

Fig. 7.26 Maps of displacements vector sum for $DLF = 1.5$ and for $t/T_1 = 0.3$ (**a**); 0.6 (**b**); 1.0 (**c**); 1.5 (**d**)

buckling (as reported earlier—Sect. 7.1), but also for the interactive dynamic buckling. On the basis of the Budiansky-Hutchinson criterion, the critical dynamic load factor, which is in the range $1.2 < DLF_{cr} < 1.5$ for the above presented example, can be determined.

During the last forty years in the world literature, many papers have been published on the interactive buckling of thin-walled structures under static load.

A wide analysis has allowed us to draw general conclusions for this interaction of various buckling modes for structures subjected to static load. However, there is a lack of such an analysis for the interactive dynamic buckling.

During the first analysis of the interactive dynamic buckling, the finite element method was used as a numerical experiment which "was telling" unexpected coupled effects of various buckling modes, never before described in the literature. To sum up, it can be concluded that the FEM package allows one, as shown above, to verify the analytical-numerical method (MAN) used for the non-linear dynamic analysis. Thus, the use of two different methods allows a peer review analysis enabling clarification of the obtained results.

References

1. Ari-Gur J, Simonetta SR (1997) Dynamic pulse buckling of rectangular composite plates. Compos B 28B:301–308
2. Hutchinson JW, Budiansky B (1966) Dynamic buckling estimates. AIAA J 4–3:525–530
3. Kolakowski Z, Kubiak T (2004) Multiple interaction of dynamic buckling modes in thin-walled members subjected in-plane pulse loading. In: Proceedings of 4th international conference on coupled instabilities in metal structure, Rome, Italy, 27–29 September 2004
4. Kolakowski Z, Kubiak T (2007) Interactive dynamic buckling of orthotropic thin-walled channels subjected to in-plane pulse loading. Compos Struct 81: 222–232
5. Kubiak T (2006) Interactive dynamic buckling of thin-walled girder with channel cross-section. In: Proceedings of international colloquium on stability and ductility of steel structures, IST Press, Lisbon, Portugal
6. Kubiak T (2007) Interakcyjne wyboczenie dynamiczne cienkosciennych slupow. Technical University of Lodz Press, Lodz
7. Volmir SA (1972) Nieliniejnaja dinamika plastinok i oboloczek. Science, Moscow
8. SAS IP, Inc (2007) ANSYS 11.1 html online documentation

Chapter 8
Girders, Beams Subjected to Pure Bending

In this chapter, the behaviour of girders or beams subjected to pulse load causing pure bending is presented. Girders with square cross-sections and beams with channel sections are considered. The part of multi-segment girders between two diaphragms is taken into account. The structures under analysis are made of isotropic or orthotropic material. The calculations were made with finite element method software and analytical-numerical method—the obtained results were compared.

The analysed girder and beams with assumed dimensions are depicted in Fig. 8.1.

All results of the numerical calculations presented in this chapter were obtained with the analytical-numerical method and the finite element method (ANSYS® software [13]). Girders with square cross-sections and of different length were considered. Elastic isotropic or orthotropic material properties were assumed in all the calculations. The material property data are presented in Table 6.1 (Chap. 6).

The analysed girders were subjected to a rectangularly shaped pulse load (Fig. 8.2) with the time duration T_p corresponding to the period of natural vibrations T of the structures under analysis.

The paper dealing with buckling analysis of beams subjected to pure bending can be easily found in word literature (e.g. [1, 2, 7–10]). However, there are lack of the papers presenting dynamic buckling of thin-walled structures subjected to pure bending.

8.1 Girders with Square Cross-Sections

It is assumed that the loaded edges of walls meet the plane section condition, and the loaded ends of girder segments are simply supported. The adopted boundary conditions and the FEM calculation models were discussed in more detail in Chap. 4 and presented in Figs. 4.11 and 4.15. The assumed model of load and the boundary conditions correspond to the analytical-numerical model [5, 6].

T. Kubiak, *Static and Dynamic Buckling of Thin-Walled Plate Structures*,
DOI: 10.1007/978-3-319-00654-3_8,

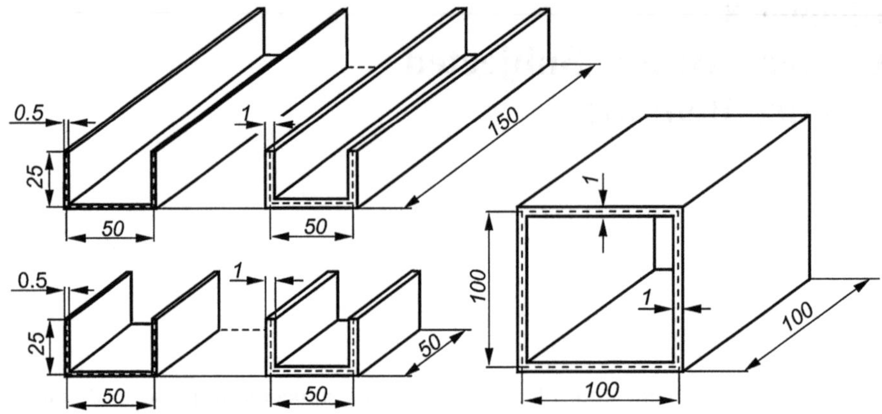

Fig. 8.1 Girder and beams under analysis

Fig. 8.2 Shape of the pulse load

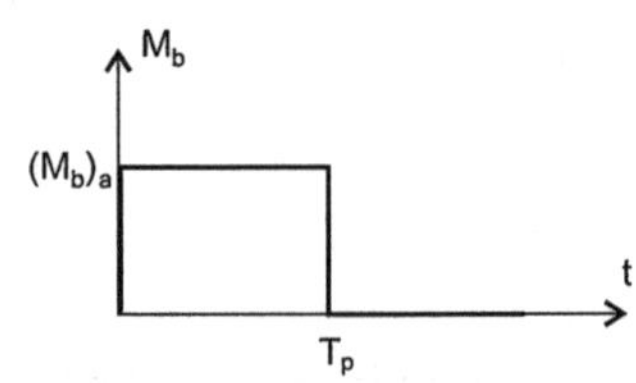

The results of calculations for the assumed and above-mentioned material property data, a type of load and boundary conditions are presented below.

The girder taken into consideration with a load distribution is presented in Fig. 8.3.

The buckling mode for girders made of a glass fibre epoxy resin composite with different volume fibre fractions f as well as girders made of steel are similar and shown Fig. 8.4. The critical moments M_{cr} for all above-mentioned cases are presented in Table 8.1.

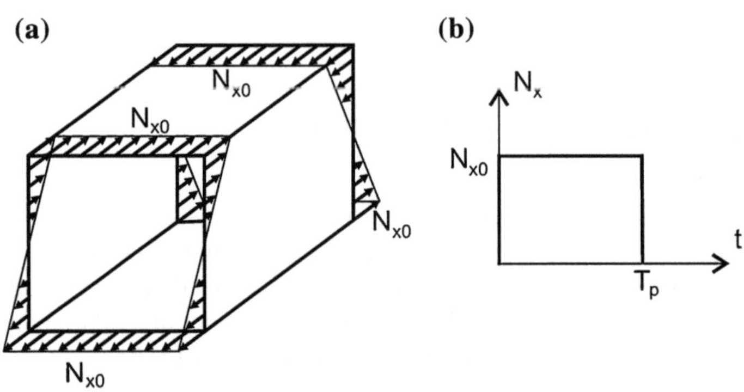

Fig. 8.3 Load distributions for the girder under analysis versus space (**a**) and time (**b**)

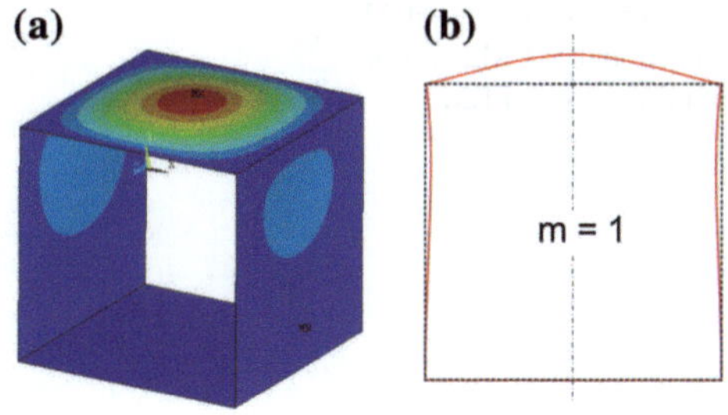

Fig. 8.4 Buckling mode obtained from FEM (**a**) and the analytical-numerical method (**b**)

Table 8.1 Buckling bending moment M_{cr} (Nm) for the girders under analysis

Method	Steel ($E_x/E_y = 1$)	Fibre glass epoxy resin		
		$f = 0.2$ ($E_x/E_y = 2.9$)	$f = 0.5$ ($E_x/E_y = 3.6$)	$f = 0.7$ ($E_x/E_y = 3.1$)
FEM	653	46.8	90.4	136.5
MAN	667	47.3	91.2	138.0
Differences[a]	2 %	1 %	>1 %	1 %

[a] Percentage difference relative to the results of the analytical-numerical method (MAN)

The results obtained with both the employed methods of calculation are almost the same. The highest differences are 2 % only. It is noted that the larger the coefficient of orthotropy, the smaller the differences in the buckling bending moment M_{cr} (see Table 8.1).

In order to determine the pulse duration, periods of natural vibrations of the considered girders were calculated. The calculations were performed for the unloaded girders and the results are summarized in Table 8.2.

As it is well known, the initial imperfection should be assumed in order to perform a dynamic buckling analysis. In the case of a compressed segment of girders with the analysed length, the fundamental modal mode and the buckling mode are the same. However, in the case of bending, there are not such similarities (Fig. 8.5). It is not possible to find the corresponding buckling and modal mode. Thus, an influence of the shape of initial imperfections on dynamic buckling was analysed. Different amplitudes of initial geometrical imperfections of girder walls and two different durations of pulse loading were analysed as well.

The results of calculations are presented in a graph showing the dimensionless deflection ξ as a function of the Dynamic Load Factor—*DLF* defined as amplitude of the applied pulse loading to the static buckling load.

Table 8.2 Frequencies and periods of natural vibrations of the girders under consideration

	Steel	Fibre glass epoxy resign		
		$f = 0.2$	$f = 0.5$	$f = 0.7$
n_1 (Hz)	482	215	271	310
T_1 (ms)	2.1	4.7	3.7	3.2

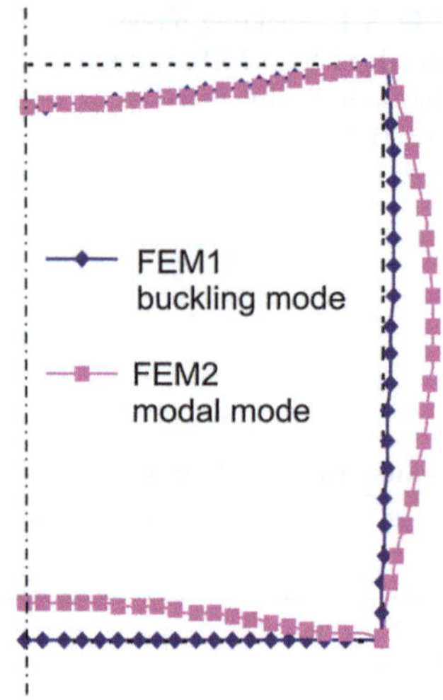

Fig. 8.5 Buckling (FEM1) and modal (FEM2) mode obtained from FEM

Analysing the dynamic response of steel girders subjected to bending pulse loading with different shapes of initial imperfections (cases FEM1 and FEM2 in Fig. 8.6), it can be seen that the shape of imperfections has a significant influence on the critical dynamic load factor DLF_{cr} obtained using the Budiansky-Hutchinson [3] or Volmir [12] criterion. For the case with the initial imperfection corresponding to the modal mode (FEM2), DLF_{cr} is in the range 1.1–1.2 but for the case FEM1 (the initial imperfection corresponds to the buckling mode), DLF_{cr} is in the range 1.7–1.8.

In Fig. 8.6, a comparison of the results obtained from FEM and the analytical-numerical method is presented (curves denoted as FEM1 and MAN, respectively). The curves obtained from the two methods used do not lie close to each other but DLF_{cr} determined from these curves using the Budiansky-Hutchinson criterion is in the same range $DLF_{cr} = 1.7$–1.8.

The next two graphs present an influence of the pulse duration and the amplitude of initial imperfections corresponding to the buckling load. The

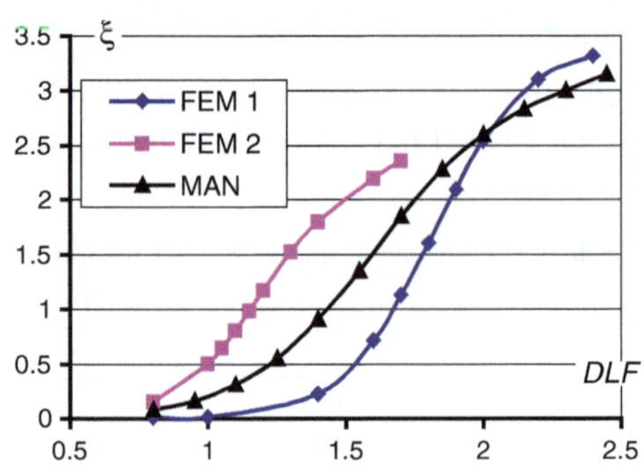

Fig. 8.6 Dimensionless deflection versus *DLF* for different shapes of initial geometrical imperfections

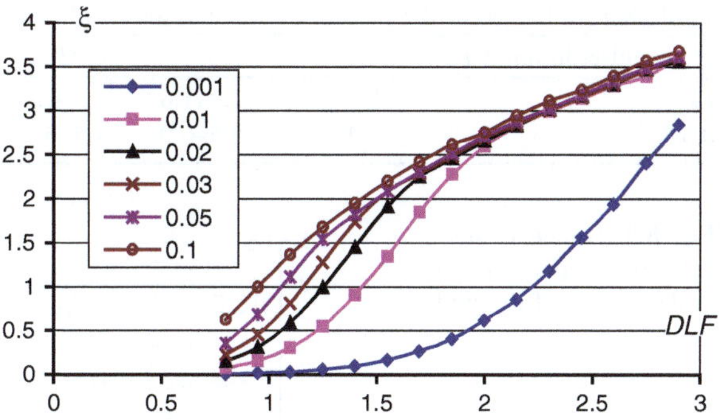

Fig. 8.7 Dimensionless deflection versus *DLF* for different amplitudes of imperfections and for the pulse duration $T_p = T$

influence of the amplitude on the critical dynamic load factor and the dynamic behaviour of steel girders subjected to pulse loading, which is pure bending, are the same as for compressed girders [4]. For higher amplitudes, the critical value of the dynamic load factor obtained from the Budiansky-Hutchinson or Volmir criterion decreases—DLF_{cr} leads to unity. The described effect does not depend on the pulse duration (compare Figs. 8.8 and 8.9).

Figures 8.7 and 8.8 present dimensionless deflection of girder walls as a function of the dynamic load factor for various amplitudes of the initial imperfections ξ^* with the shape corresponding to the buckling mode. Comparing the results presented in Figs. 8.7 and 8.8, one can see an influence of the pulse duration. For the pulse duration equal half a period of natural vibrations, DLF_{cr}s are higher than for the pulse duration corresponding to the period of natural vibrations. In Table 8.3, values of DLF_{cr} obtained from the curves presented in Figs. 8.7 and 8.8 using the Budiansky-Hutchinson criterion are given together.

Let us compare an influence of material properties on the dynamic response and the critical value of the dynamic load factor DLF_{cr}. Dimensionless deflection of girder walls as a function of the dynamic load factor for a steel girder and composite girders with different volume fibre fractions is presented in Fig. 8.9. The results of calculations were obtained using FEM with the assumed time of pulse duration equal to a period of the natural frequency $T_p = T_1$ and a shape of the

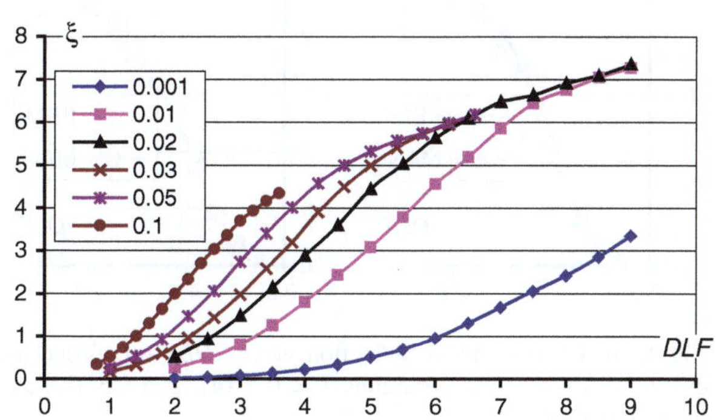

Fig. 8.8 Dimensionless deflection versus *DLF* for different amplitudes of imperfections and for the pulse duration $T_p = 0.5$ T

Table 8.3 Influence of the pulse duration and the amplitude of initial imperfections on DLF_{cr}

ξ^* = pulse duration T_p=	0.001	0.01	0.02	0.03	0.05	0.1
T_1	2.4–2.5	1.7–1.8	1.3–1.5	1.1–1.2	0.9–1.1	0.9–1
$0.5T_1$	>9	5.5–6.0	4.5–5	3.8–4.2	2.6–3.0	2.2–2.4

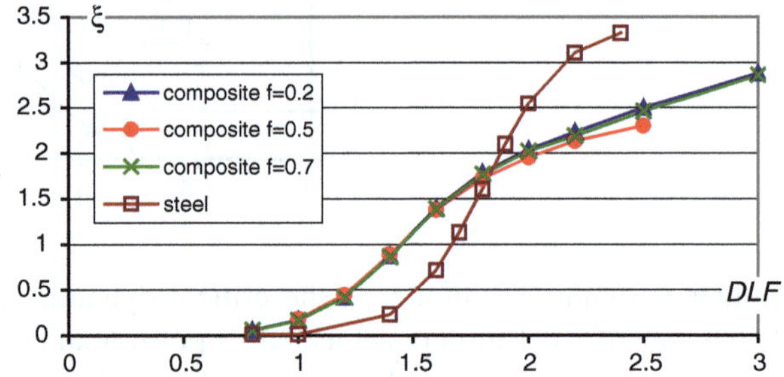

Fig. 8.9 Dimensionless deflection versus *DLF* for girders made of different materials

initial imperfection corresponding to the buckling mode with an amplitude equal 1/100 thickness of the girder wall. A comparison between the results from FEM and MAN (analytical-numerical method) are presented in Fig. 8.10.

Comparing the course of the curves in Fig. 8.9, one can see that the coefficient of orthotropy exerts the main influence on the dynamic response. The curves for composite girders with the volume fibre fraction $f = 0.2$ and $f = 0.7$ are almost identical—the coefficients of orthotropy E_x/E_y for the above-mentioned materials are equal to 2.9 and 3.1. The curve denoted by "composite $f = 0.5$" has a slightly different course than the one previously mentioned—for a composite with $f = 0.5$, the coefficient of orthotropy equals 3.6. The relation between the dimensionless deflection as a function of the dynamic load factor for steel girders is completely different than for composite girders because these girders are made of an isotropic material with the coefficient of orthotropy E_x/E_y equal to 1.

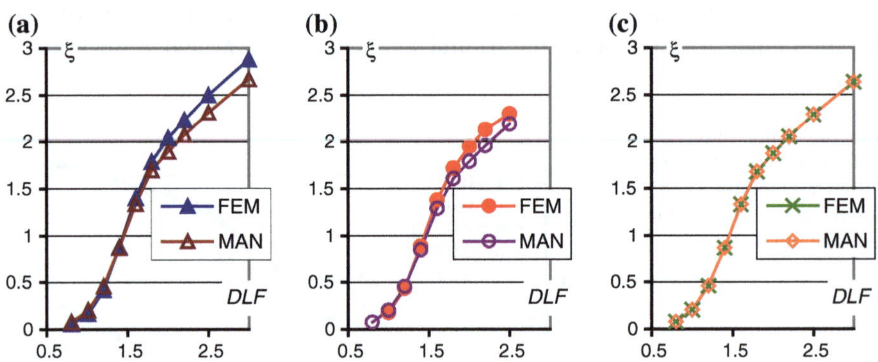

Fig. 8.10 Dimensionless deflection versus *DLF* for girders made of composite materials with different volume fibre fractions f = 0.2 (**a**), f = 0.5 (**b**) and f = 0.7 (**c**)—a comparison of numerical methods

Comparing the results for composite girders with different volume fibre fractions obtained with FEM and MAN, one can see that the critical dynamic load factor according to the Budiansky-Hutchinson or Volmir criterion is exactly the same. The differences appear for a higher *DLF* value. The deflections obtained from FEM are slightly higher than those obtained from the analytical-numerical method. Such behaviour can be explained by stiffness of the model—the analytical-numerical model is stiffer than the finite element model which has many more degrees of freedom.

8.2 Channel Section Beams

The calculations were performed with FEM only. It is assumed that the loaded edges of walls meet the plane section condition, and the loaded ends of girder segments are simply supported [11]. The adopted boundary conditions and the FEM calculation model were discussed in more detail in Chap. 4 and presented in Fig. 4.16 (case BC-2).

Channel cross-section beams are made of steel with the material property data given in Table 6.1. The assumed load cases are presented in Fig. 8.11.

Pure bending in the plane normal to the web of the channel section beam is denoted by case 1 (see Fig. 8.11a)—the web is compressed and the flanges are

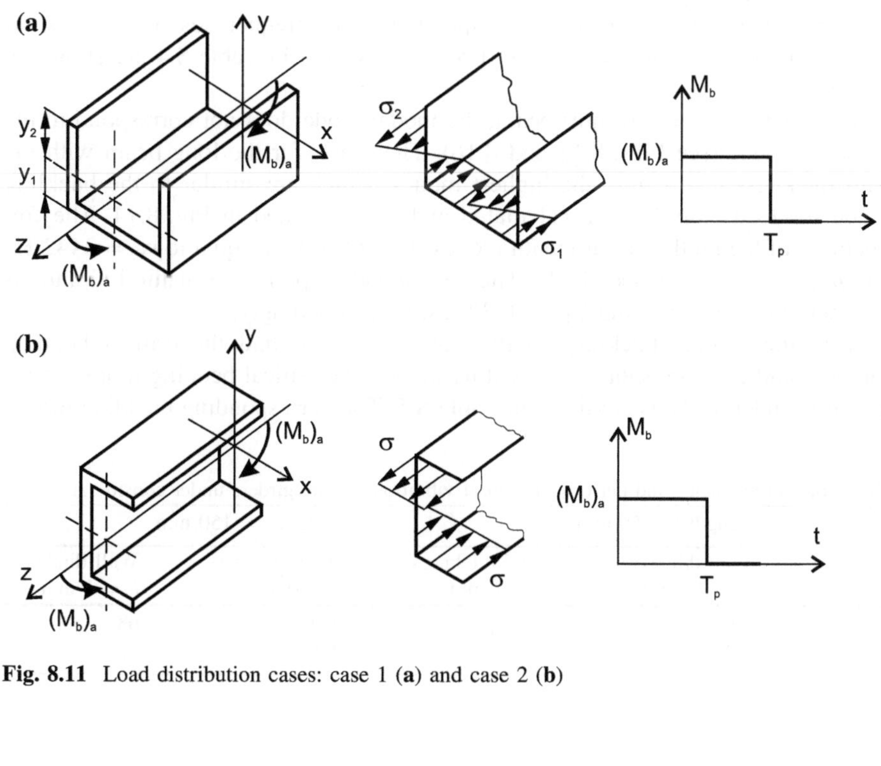

Fig. 8.11 Load distribution cases: case 1 (**a**) and case 2 (**b**)

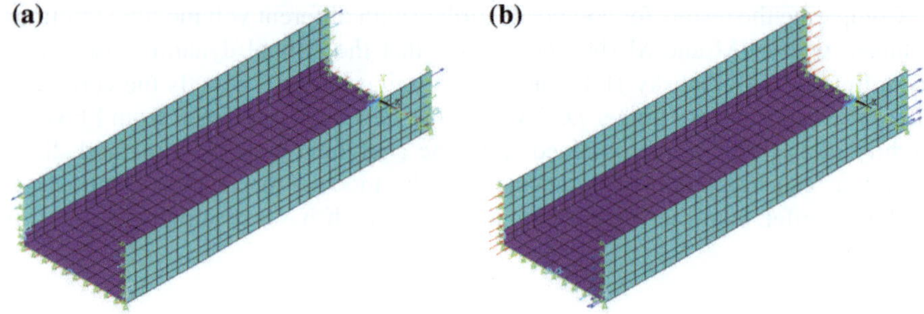

Fig. 8.12 Finite element models for load case 1 (**a**) and load case 2 (**b**)

under plane bending. Pure bending in the plane parallel to the web of the considered channel section beam is denoted by case 2 (Fig. 8.11b)—the web is under plane bending, the upper flange is under tension, whereas the bottom flange is compressed. The finite element models for both the load cases taken into account are presented in Fig. 8.12.

For load case 1, the calculations were performed for two thicknesses of the beam walls $h = 1$ mm and $h = 0.5$ mm (see Fig. 8.1). The calculations for load case 2 were performed only for the beam with the wall thickness $h = 1$ mm.

The modal analysis was performed to find frequencies of first fundamental natural vibrations for all channel beams taken into account. The frequencies and the corresponding periods of natural vibrations are presented in Table 8.4.

The modal modes which correspond to the frequencies for the short ($l = 50$ mm) and long ($l = 150$ mm) beam presented in Table 8.4 are shown in Fig. 8.13.

For the long ($l = 150$ mm) beam, the modal mode does not correspond to the buckling mode (see Figs. 8.15 and 8.16). Therefore, for the long beam with the wall thickness $h = 1$ mm, the highest modes, which are similar to the buckling mode for load case 1, were calculated and are presented in Fig. 8.14. The frequencies of the ninth and thirteenth modes (Fig. 8.14a) are equal to $n_9 = 2442$ Hz and $n_{13} = 3161$ Hz, respectively. The corresponding periods of natural vibrations are then $T_9 = 0.41$ ms and $T_{13} = 0.32$ ms, correspondingly.

The linear static buckling analysis allows one to find the critical bending moment and the corresponding buckling mode. The critical bending moments for all cases under analysis are listed in Table 8.5. The corresponding buckling modes

Table 8.4 Frequencies and periods of natural vibrations of the girders under consideration

	Length $l = 50$ mm		Length $l = 150$ mm	
	Wall thickness $h = 0.5$ mm	Wall thickness $h = 1$ mm	Wall thickness $h = 0.5$ mm	Wall thickness $h = 1$ mm
n_1 (Hz)	836	1666	344	685
T_1 (ms)	1.2	0.6	2.9	1.5

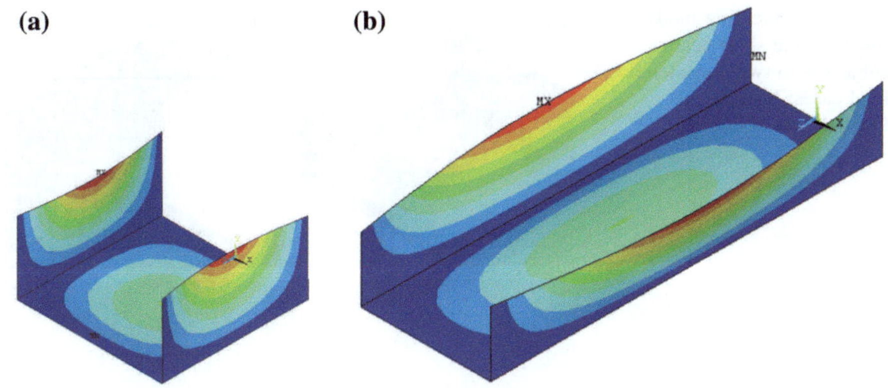

Fig. 8.13 First modes of natural vibrations for a C-shape beam of the length $l = 50$ mm (**a**) and $l = 150$ mm (**b**)

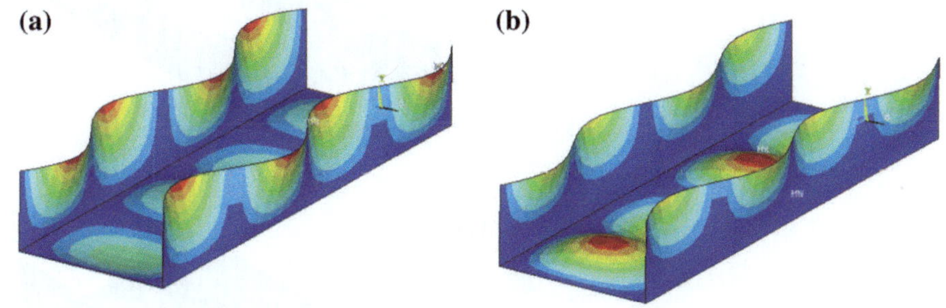

Fig. 8.14 Ninth (**a**) and thirteenth (**b**) modal mode

are shown for the short ($l = 50$ mm) beam in Fig. 8.15 and for the long ($l = 150$ mm) beam in Fig. 8.16.

First, the results of the dynamic response of short ($l = 50$ mm) beams subjected to load case 1 and load case 2 are discussed. The time of pulse duration was equal to the period of natural vibrations T_1 (see Table 8.4). The results are presented as a relation between the dimensionless deflection $\xi = w/h$ versus the dynamic load

Table 8.5 Critical bending moment M_{cr} (Nm) for the analysed channel section beams

	Length $l = 50$ mm		Length $l = 150$ mm	
	Wall thickness $h = 0.5$ mm	Wall thickness $h = 1$ mm	Wall thickness $h = 0.5$ mm	Wall thickness $h = 1$ mm
Load case 1	57 ($m = 1$)[a]	449 ($m = 1$)	54 ($m = 4$)	427 ($m = 4$) 1565 ($m = 1$)
Load case 2	–	440 ($m = 1$)	–	442 ($m = 3$)

[a] m is a number of halfwaves along the beam

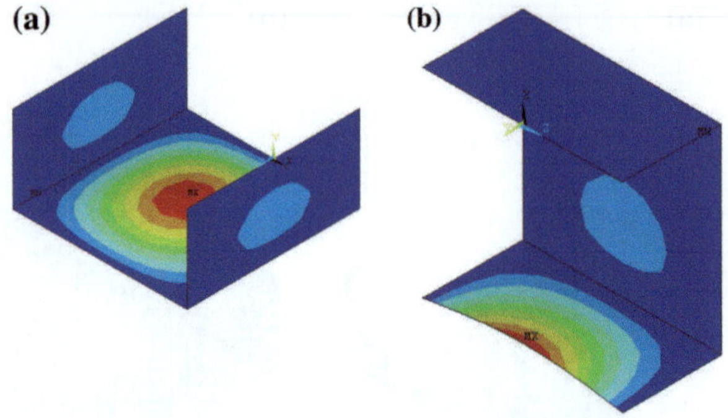

Fig. 8.15 Buckling mode m=1 for the short beams subjected to two considered load cases: load case 1 (**a**) and load case 2 (**b**)

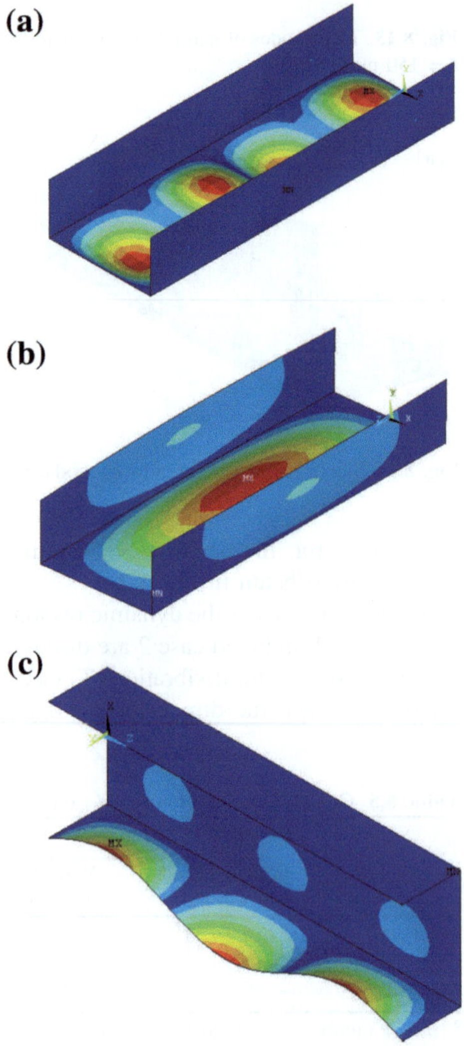

Fig. 8.16 Buckling mode for the medium length beams subjected to two considered load cases: load case 1 (**a**) and load case 2 (**b**)

factor *DLF*. The initial imperfection corresponding to the buckling mode with the amplitude equal to 1/100 thickness of the beam wall was assumed. The dynamic response for channel beams with two different wall thicknesses *h* subjected to bending pulse loading are presented in Fig. 8.17 for load case 1 and in Fig. 8.18 for load case 2, respectively.

In both the assumed load cases and both the wall thicknesses, the courses of curves ξ (*DLF*) suggest that a local loss of stability occurred in all the above-mentioned cases. The critical value of dynamic load factors DLF_{cr} was determined according the Volmir and Budiansky-Hutchinson criteria. The critical value of dynamic load factors DLF_{cr} for channel section beams subjected to bending are less than for the same structures subjected to uniform compression. Using the Volmir criterion, DLF_{cr}s are approximately equal to 0.48 and 0.88 for beams with the wall thickness $h = 0.5$ mm and $h = 1$ mm, respectively. The value of DLF_{cr} according to the Budiansky-Hutchinson criterion for beams with two above-mentioned wall thicknesses are $DLF_{cr} = 0.45$–0.5 for $h = 0.5$ mm and $DLF_{cr} = 1.0$–1.2 for h = 1 mm. Such a low value of the critical dynamic load factor could be explained by very low bending stiffness for channel section beams subjected to bending in the plane perpendicular to the beam web. Taking into account load case 2, the channel section beam is subjected to bending in plane with its highest moment of inertia. Changing the plane of load leads to an

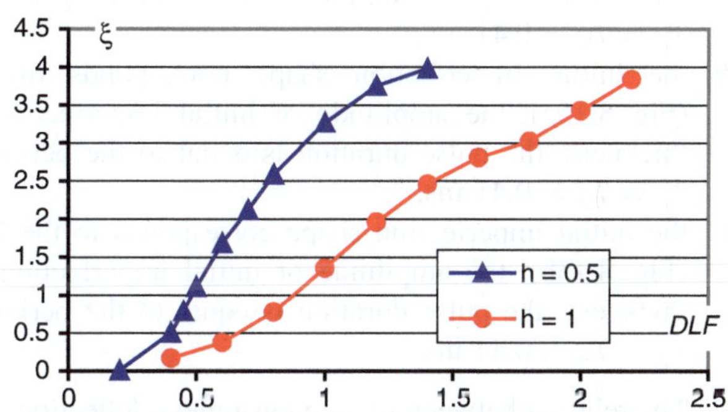

Fig. 8.17 Dynamic response for channel section beams with different wall thickness *h*—load case 1

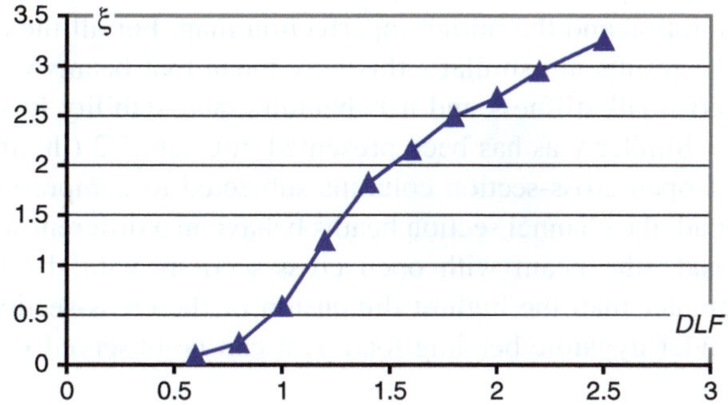

Fig. 8.18 Dynamic response for the channel section beam with the length $l = 50$ mm and the wall thickness $h = 1$ mm—load case 2

inconsiderable increase in DLF_{cr}, which according to the Volmir criterion is equal 1.15 and according to the Budiansky-Hutchinson criterion, is in the range 1.1–1.2 (Fig. 8.18).

Completely different results were obtained for beams with the length $l = 150$ mm and the wall thickness $h = 1$ mm subjected to the pulse bending moment—load case 1. The results are shown in Fig. 8.18 and the curves are denoted as follows:

1. the initial imperfection shape corresponds to the first modal mode (Fig. 8.13b), the amplitude of initial imperfections equals 1/100 of the wall thickness, the pulse duration is equal to the period of the first modal mode $T_p = T_1 = 1.5$ ms (Table 8.4);
2. the initial imperfection shape corresponds to the thirteenth modal mode (Fig. 8.14b), the amplitude of initial imperfections equals 1/100 of the wall thickness, the pulse duration is equal to the period of the thirteenth modal mode $T_p = T_{13} = 0.32$ ms;
3. the initial imperfection shape corresponds to the ninth modal mode (Fig. 8.14a), the amplitude of initial imperfections equals 1/100 of the wall thickness, the pulse duration is equal to the period of the ninth modal mode $T_p = T_9 = 0.41$ ms;
4. the initial imperfection shape corresponds to the local buckling mode (Fig. 8.15a), the amplitude of initial imperfection equals 1/100 of the wall thickness, the pulse duration is equal to the period of the ninth modal mode $T_p = T_9 = 0.41$ ms;
5. the initial imperfection shape corresponds to the local buckling mode (Fig. 8.15a), the amplitude of initial imperfections equals 1/10 of the wall thickness, the pulse duration is equal to the period of the ninth modal mode $T_p = T_9 = 0.41$ ms;
6. the initial imperfection shape corresponds to the buckling mode with m = 1 (Fig. 8.15b), the amplitude of initial imperfections equals 1/100 of the wall thickness, the pulse duration is equal to the period of the ninth modal mode $T_p = T_9 = 0.41$ ms.

The relation between the dimensionless deflection and the dynamic load factor is almost linear, which could mean that the buckling phenomena did not occur. The calculations were repeated with different assumptions for the period of pulse duration and the initial imperfection map. For all the cases presented in Fig. 8.19, the results are similar—this may mean that beams with open cross-sections have too small stiffness and it is bending, and stability loss does not occur.

Similarly as has been presented in Chap. 7.2 (the interactive dynamic buckling of open cross-section columns subjected to compression), in the case of bending load, the channel section beams behave in a different way under static and dynamic load—the beam with open cross-sections with the length three or more times greater than the highest dimension of the cross-section does not lose its stability under dynamic bending load as it can be observed if load is static.

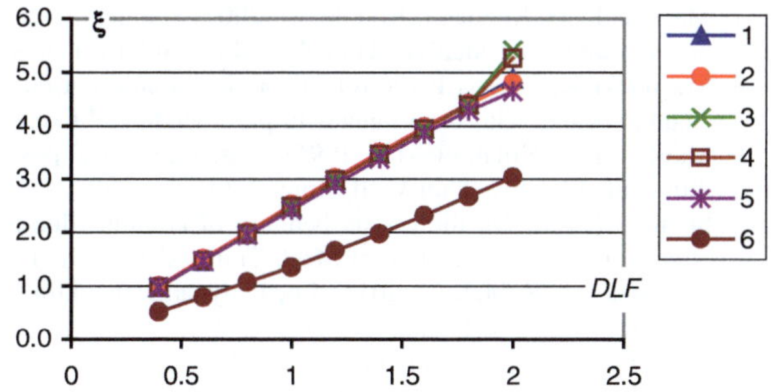

Fig. 8.19 Dynamic response for the channel section beam with the length l = 150 mm—load case 1

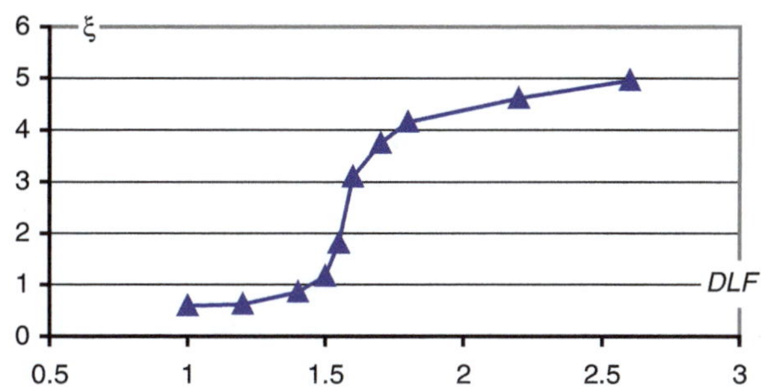

Fig. 8.20 Dynamic response for the channel section beam with the length l = 150 mm and the wall thickness h = 1 mm—load case 2

After rotating the cross-section (load case 2), the stiffness of the beam grows and the dynamic buckling phenomena can be observed (Fig. 8.20).

Using the course of the curve presented in Fig. 8.20, it is possible to determine the critical value of the dynamic load factor, which is equal to 1.45 according to the Volmir criterion and is in the range 1.55–1.6 according to the Budiansky-Hutchinson criterion.

References

1. Graves-Smith TR (1969) The local buckling of box girders under bending stresses. Int J Mech Sci 11:603–612
2. Graves-Smith TR (1972) The postbuckled behaviour of a thin-walled box beam in pure bending. Int J Mech Sci 14:711–722
3. Hutchinson JW, Budiansky B (1966) Dynamic buckling estimates. AIAA J 4–3:525–530
4. Kowal-Michalska K, Mania R (2005) Dynamic buckling load of thin-walled columns of closed cross-section subjected to pulse loading. In: Proceedings of SSTA, AA Balkema Publisher, Rotterdam, pp 353–358
5. Krolak M, Mania RJ (eds) (2011) Static, dynamic and stability of structures. Vol. 1: stability of thin-walled plate structures. A series of monographs. Technical University of Lodz Press, Lodz
6. Kubiak T (2007) Metoda elementow skonczonych jako eksperyment numeryczny statecznosci dzwigarow cienkosciennych obciazonych statycznie i dynamicznie. In: Niezgoda T (ed) Numerical analysis of selected problem in mechanics. Military University of Technology Press, Warsaw, pp 209–228

7. Macdonald M, Rhodes J, Kotelko M (2007) Stainless steel stub columns subject to combined bending and axial loading. Thin Wall Struct 45(10–11):893–897
8. Magnucki K, Paczos P (2009) Theoretical shape optimization of cold-formed thin-walled channel beams with drop flanges in pure bending. J Constr Steel Res 65(8–9):1731–1737
9. Manevich AI, Kolakowski Z (1996) Influence of local postbuckling behaviour on bending of thin-walled beams. Thin Wall Struct 25(3):219–230
10. Sevin E (1960) On the elastic bending of columns due to dynamic axial forces including effects of axial inertia. J Appl Mech Trans ASME 27(1):125–131
11. Urbaniak M, Kubiak T (2011) Local dynamic buckling of C-shape profiles subjected to bending. Mech Mech Eng 15(1):129–144
12. Volmir SA (1972) Nieliniejnaja dinamika plastinok i oboloczek. Science, Moscow
13. SAS IP, Inc (2007) ANSYS 11.1 html online documentation

Chapter 9
Thin-Walled Girders Subjected to Torsion

The results of the dynamic buckling analysis of girders subjected to torsion are presented in this chapter. All the results of numerical calculations to be discussed further were obtained with finite element method commercial software. Girders with square cross-sections and different lengths were considered. A little attention is paid to thin-walled structures with flat wall subjected to bending or torsion. It was therefore decided to fill this gap by conducting an analysis of thin-walled girders with closed cross-section and behaviour of such structures subjected to torsion impulse of finite duration.

The assumed dimensions of the analysed girders are presented in Fig. 9.1. Three different girders length l were taken into account ($l = 100$, 200 and 400 mm). All calculations were performed on the assumption of elastic isotropic material properties, which are as follows:

- Young's modulus: $E = 2{\cdot}10^5$ MPa;
- Poisson's ratio: $\nu = 0.3$;
- density: $\rho = 7853$ kg/m^3.

The influence of material damping was also considered.

The Ansys® software [7] has been employed to model and analyse dynamic behaviour of thin-walled girders.

To introduce the load causing twist, one end of the girder was restrained (for all nodes of the FEM model, three perpendicular displacements were set to zero) and two pairs of forces were applied (Fig. 4.22) to the second end, creating a resultant torque. The assumed boundary conditions were the same as presented in Fig. 4.24.

It was assumed that the girders were subjected to pulse load of a rectangular shape (Fig. 9.2). The duration time of the pulse torque T_p corresponds to the fundamental natural vibration period $T = 1/f$, where f is the first natural frequency. Table 9.1 shows the results of eigenvalue calculations (i.e., the frequency and the period of natural vibrations) for different lengths of the girders considered.

Linear stability analyses which determined the critical load and the corresponding buckling mode were carried out as well. The critical values of torque that

T. Kubiak, *Static and Dynamic Buckling of Thin-Walled Plate Structures*,
DOI: 10.1007/978-3-319-00654-3_9,

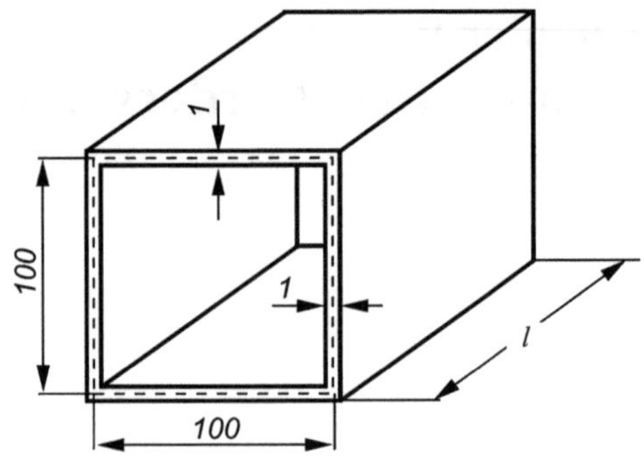

Fig. 9.1 Girders under analysis

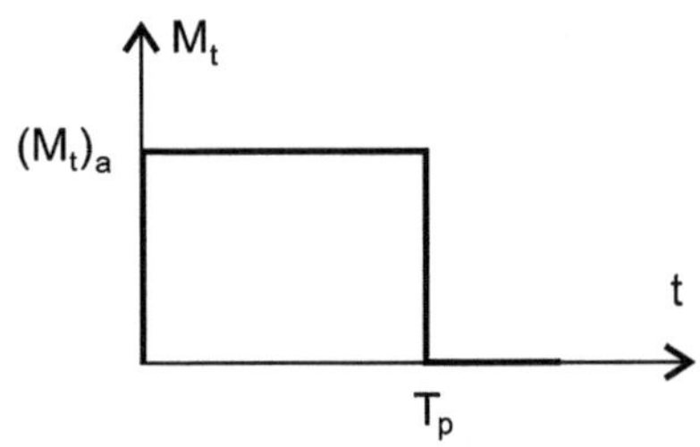

Fig. 9.2 Shape of the torsional pulse loading

Table 9.1 Frequency and the period of natural vibrations for the girders under analysis

Girder length l (mm)	100	200	400
Frequency of natural vibrations f (Hz)	574	314	257
Period of natural vibrations T (ms)	1.7	3.2	3.9

were taken to determine the *DLF* value for each analysed girders are shown in Table 9.2.

In the dynamic buckling analysis, the initial geometrical imperfection should be assumed [5]. For compressed structures, the assumed shape of initial imperfection corresponds to the buckling mode which is identical with the modal mode of the frequency taken to determine the pulse duration. In the case of structures subjected to torque, there is no possibility to find the modal mode (with the independent vibration frequency) corresponding to the buckling mode of girders subjected to torsion (Fig. 9.3).

Table 9.2 Critical torque for the girders under analysis

Girder length l (mm)	100	200	400
Critical torque $(M_t)_{cr}$ (Nm)	4.313	2.576	2.105

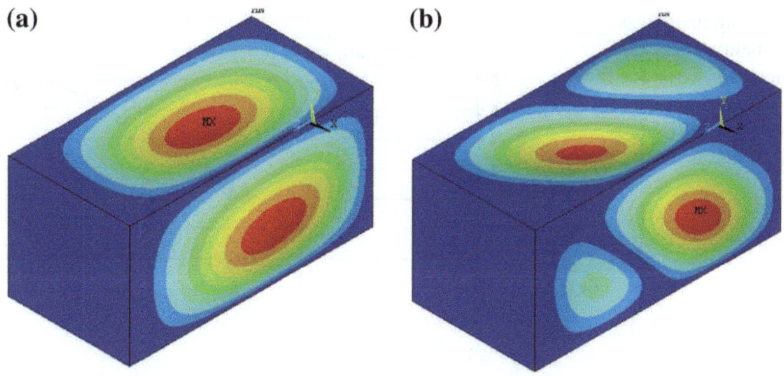

Fig. 9.3 Modal (**a**) and buckling (**b**) mode in the girder subjected to torsion

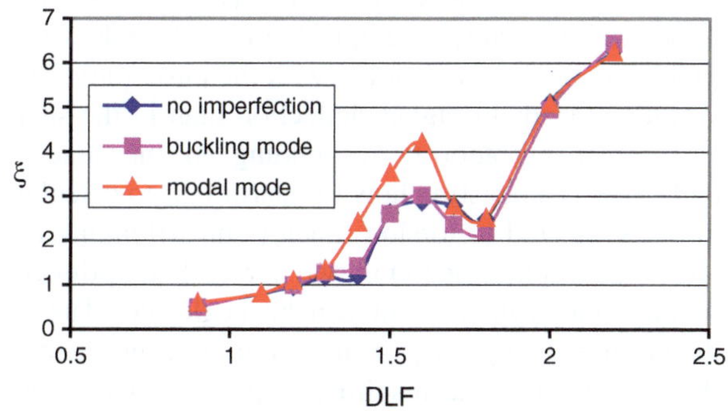

Fig. 9.4 Dimensionless displacement versus the dynamic load factor for the girder with different shapes of initial imperfections

Let us see what an influence of the shape and the amplitude ($w^* = w_0/h$, where w_0 is the assumed initial deflection and h is the girder wall thickness) of initial geometrical imperfections on the dynamic response of the girder subjected to pulse torsion is. In Fig. 9.4, dimensionless deflection versus the dynamic load factor for the girder with the length $l = 200$ mm and different shapes (corresponding to the modal mode or to the buckling mode) of geometrical imperfections with the amplitude $w^* = 0.01$ is presented.

In Fig. 9.5, dimensionless deflection versus the dynamic load factor for the girder with the length $l = 100$ mm and different amplitudes w^* of geometrical imperfections with the shape corresponding to the buckling mode is presented.

Analysing the results presented in Figs. 9.4 and 9.5, one can say that the amplitude of initial imperfections has almost no impact on the course of the curves. It means that the amplitude of initial imperfections with the shape corresponding to the buckling mode does not influence the critical value of the dynamic load factor. The results of the analysis presented further were obtained on the assumption of no initial geometrical imperfections.

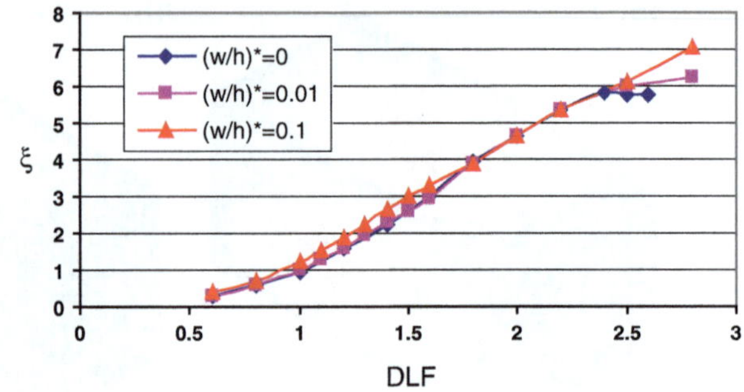

Fig. 9.5 Dimensionless displacement versus the dynamic load factor for the girder with different amplitudes of initial imperfections

Nevertheless, according to the results presented in Fig. 9.4, it should be noted that assuming the initial imperfection shape corresponding to the modal mode, higher deflection is obtained than for the girder with initial imperfections corresponding to the buckling mode. To conclude, it can be said that the initial imperfection corresponding to the modal mode assumed for the girders subjected to the torsional pulse loading with the pulse duration equal to the period of natural vibrations leads to higher deflection than in the shape of initial imperfections.

Another problem corresponding to the load case and the corresponding deflection is a question how to measure or to calculate deflection of the wall of the girder subjected to torsion. Each point belonging to the thin-walled girder translates (has displacements) in two perpendicular directions—if it is assumed that the longitudinal displacement can be neglected. The first possibility is to take the displacements u_{Ox}, u_{Oy} of the given point in x and y directions, respectively, and then to calculate the resulting displacement w_{abs} of the given point in the plane perpendicular to the girder axis according to the following formula [4]:

$$w_{abs} = \sqrt{u_{Ox}^2 + u_{Oy}^2}. \tag{9.1}$$

The next possibility is to separate the displacement corresponding to twist from those corresponding to buckling. The proposed solution is based on an analysis of movements of three nodes located in the plane of the selected cross-section of the girder in a single line on the same wall (from edge to edge). The value of the dynamic deflection w corresponding to a loss of stability (to enlarge the initial deflection) is determined from the following formula [4]:

$$w = \sqrt{\left(u_{Ox} - u_{av\_x}\right)^2 + \left(u_{Oy} - u_{av\_y}\right)^2}. \tag{9.2}$$

where u_{Ox}, u_{Oy} are two perpendicular displacements of point O (Fig. 9.6) and the average displacements u_{av_x}, u_{av_y} are calculated as follows:

$$u_{av\_x} = \frac{1}{2}\left(u_{Lx} + u_{Rx}\right), \quad u_{av\_y} = \frac{1}{2}\left(u_{Ly} + u_{Ry}\right). \tag{9.3}$$

where u_{Lx}, u_{Ly} and u_{Rx}, u_{Ry} are two perpendicular displacements of points L and R lying on the ends of the selected line (Fig. 9.6).

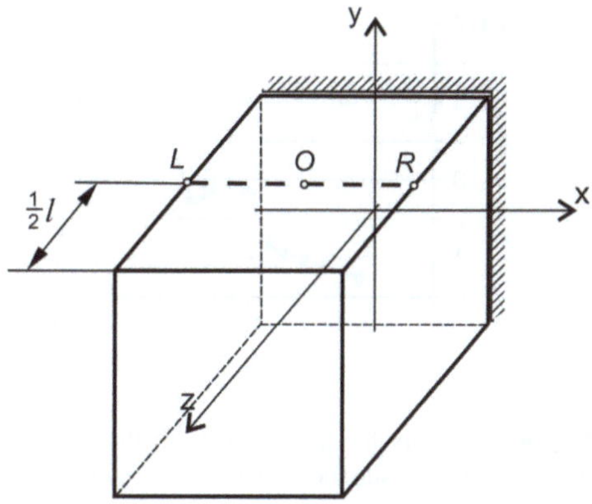

Fig. 9.6 Points taken to calculate the deflection corresponding to buckling

To check an influence of the way of collecting deflections on results of the dynamic response, four different cases were taken into account. In all the cases, the displacement was measured in the middle of the length and the width of the upper wall of the girder. The following notation was used:

- w_{imp} presents the maximal displacement obtained during the pulse duration ($0 < t < T_p$) for each applied *DLF* and calculated using (9.2);
- w—displacement calculated using (9.2) but the maximum value was taken from the whole period of tracking the dynamic impulse response ($0 < t < 1.5T_p$);
- w_{abs_imp}—normal maximal displacement to the girder wall without any recalculation, obtained during the pulse duration ($0 < t < T_p$) for each applied *DLF;*
- w_{abs}—normal maximal displacement to the girder wall without any recalculation, taken from the whole period of tracking the dynamic impulse response ($0 < t < 1.5T_p$).

An influence of different measurements way of deflection on the dynamic response is presented in Fig. 9.7, which shows dimensional deflection versus the dynamic load factor for the 200 mm length girder subjected to pulse torsion. The dynamic load factor was calculated dividing the pulse amplitude $(M_t)_a$ by the static critical torque $(M_t)_{cr}$.

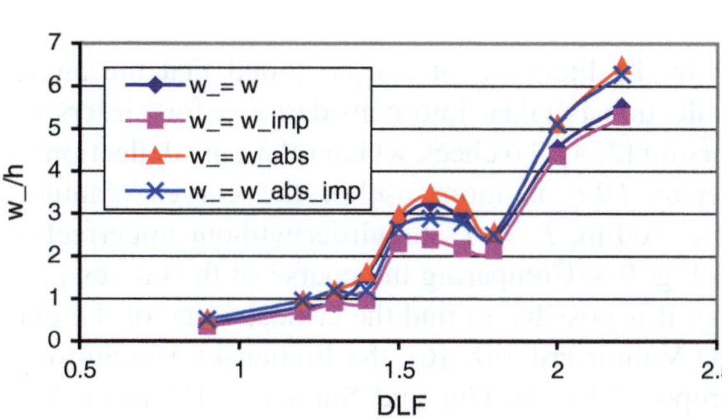

Fig. 9.7 Different representations of the dimensionless displacement versus the dynamic load factor for the girder with $l = 200$ mm

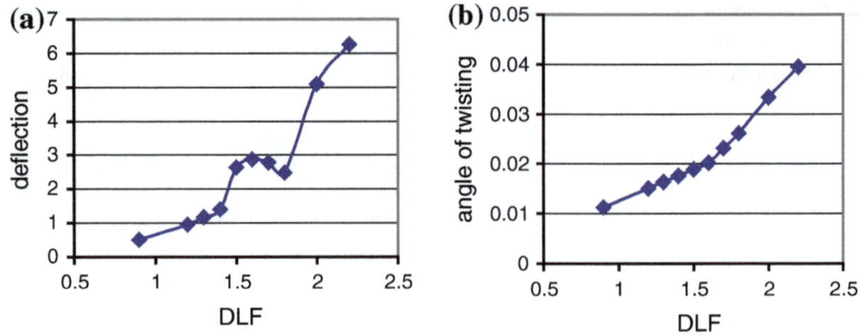

Fig. 9.8 Results for the girder with the length $l = 200$ mm presented as **(a)** a dimensionless displacement versus the dynamic load factor. **b** an angle of twist versus the dynamic load factor

Fig. 9.9 Different representations of the dimensionless displacement versus the dynamic load factor for the girder with $l = 200$ mm and the material damping ratio equal to 2 %

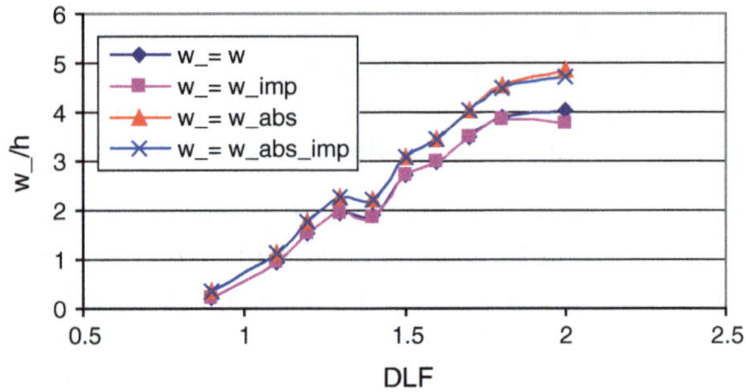

Fig. 9.10 Different representations of the dimensionless displacement versus the dynamic load factor for the girder with $l = 200$ mm and the material damping ratio equal to 5 %

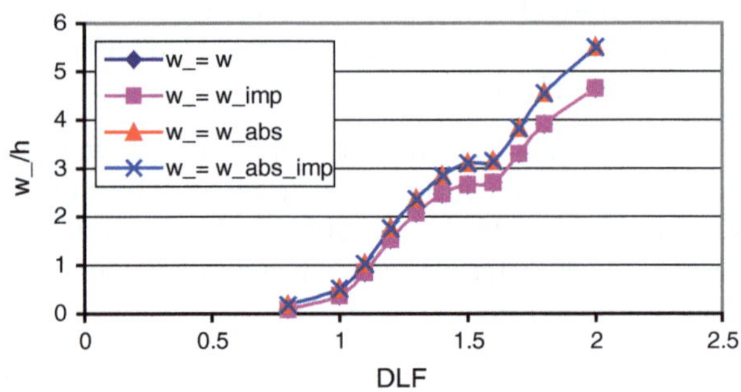

In the literature, it can be found that the angle of twist instead of the wall deflection is taken into consideration for girders or other structures subjected to torsion [2, 4]. To check which relations: deflection versus *DLF* or an angle of twist versus *DLF* is more useful, the curves obtained for exactly the same case ($l = 200$ m, $T_p = T$, the girder without imperfection) are prepared and presented in Fig. 9.8. Comparing the course of the curves presented in Fig. 9.8, one can see that it is possible to find the critical value of the dynamic load factor according to the Volmir criterion [6], the Budiansky-Hutchinson criterion [3] and the criterion proposed by Ari-Gur and Simonetta [1] from the curve representing a relation

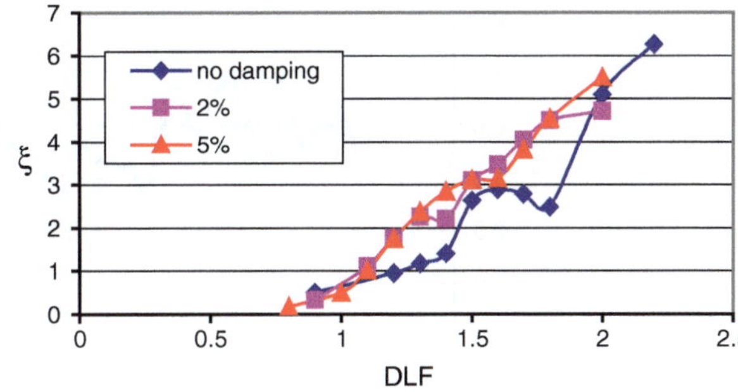

Fig. 9.11 Different representations of the dimensionless displacement versus the dynamic load factor for the girder with $l = 200$ mm and the damping ratio equals 5 %

Table 9.3 Critical value of *DLF* obtained with different criteria

	Volmir	Budiansky-Hutchinson	Ari-Gur and Simonetta
Undamped	1.2	1.4 ÷ 1.5	1.6
Damping ratio 2 %	1.1	1.2 ÷ 1.3	1.3
Damping ratio 5 %	1.1	1.2 ÷ 1.3	1.5

between the girder wall deflection and the dynamic load factor. From the curve representing an angle of twist as a function of *DLF*, only the Budiansky-Hutchinson criterion can be used. To summarise the above, deflection as a function of *DLF* is more useful to apply different criteria of dynamic buckling.

Figures 9.9 and 9.10 present the same curves as in Fig. 9.8 obtained for the same girder but with arbitrary assumed material damping equal to 2 % and 5 %, respectively.

Considering the results of calculations presented in Figs. 9.7, 9.8 and 9.10, it could be said that it will be enough to take the curves denoted by $w\_ = w\_abs\_imp$ to analyse dynamic buckling and determine the critical amplitude of the pulse load (DLF_{cr}).

To analyse the damping influence, the relation w_{abs_imp}/h (denoted w/h) versus *DLF* was chosen, as suggested above. The results for different damping ratios are presented in Fig. 9.11. Comparing these results, it can be said that in contrast to the compressed girders, damping has an influence on the critical value of DLF_{cr}. For damped girders, DLF_{cr} is lower than for the undamped ones. Values of DLF_{cr} for damped and undamped girders obtained with the above-mentioned criteria are presented in Table 9.3.

Analysing the results presented in Fig. 9.11, one can notice that the displacements for the damped girder are higher than for the undamped one. Such a relationship seems to be unpredictable. To explain it, let us see the time dependent displacement curves (Fig. 9.12) and the girder displacement map (Fig. 9.13) for $DLF = 1.4$. The girder displacement map has been prepared for the time at which the displacements achieve their maximum value.

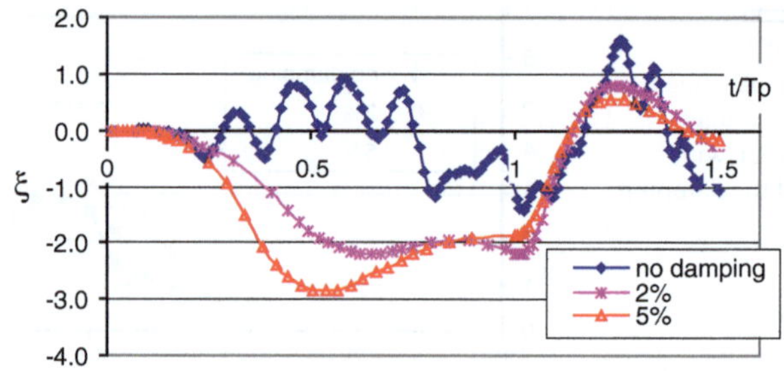

Fig. 9.12 Time dependent displacement measured in the middle of the length and the width on the upper wall of the girder for $DLF = 1.4$

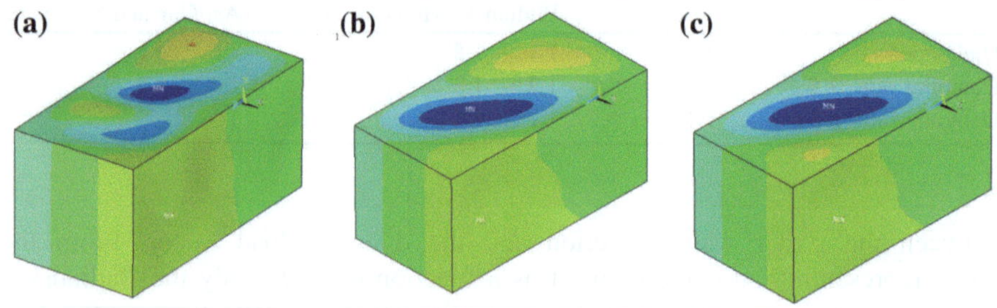

Fig. 9.13 Normal to upper wall displacement map for the undamped (**a**) and damped girder with the ratio 2 % (**b**) and 5 % (**c**)

Comparing the time dependent displacement and the mode (or the deflection shape) of the upper wall of girders under analysis, one can say that damping exerts an influence on the girder dynamic behaviour in the case of torsion. First of all, completely different shapes of deflection appear for damped and undamped girders. As can be seen for the undamped case, about three halfwaves appear in the longitudinal direction and about two halfwaves in the transverse direction on the upper wall of the analysed girder. In both the damped cases, two halfwaves appear: one low and one high (the amplitudes differ approximately three times). Concluding, it can be said that the previously unpredictable differences in the displacements are the results of completely different shapes of displacements. If the obtained displacement is multiplied by a number of halfwaves to recalculate the displacement for one halfwave only, the results of the maximum displacement will be as follows: $w = 1.45 \times 3 = 4.35$ for the undamped girder and $w = 2.86 \times 1.33 = 3.8$. Now the relations look correct. The displacements are higher for the undamped girder than for the damped one.

The same way as the above-described one can be used to explain the difference in dynamic responses between the girder with the damping ratio equal to 2 % and 5 %. In addition, the maximum deflection of the girder with the damping ratio 5 % lies closer to the reference measurement point than for the girder with the damping ratio of 2 %.

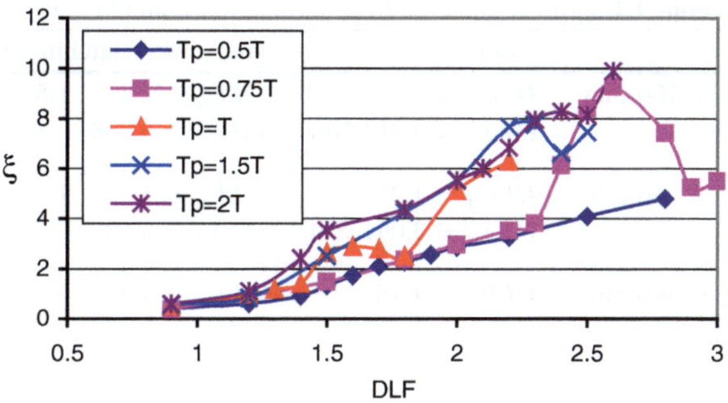

Fig. 9.14 Influence of the pulse duration on the dimensionless displacement versus the dynamic load factor for the girder with the length $l = 200$ mm

To conclude the above considerations, it should be said that the damping ratio has a great influence on the results of calculations for the girder subjected to the torsional pulse loading. Nevertheless, the true damping ratio value is dependent on many parameters and is impossible to be designated precisely. Therefore, the author has decided to neglect damping in the further calculations checking an influence of time of pulse duration and the length of girders on differences in the critical value of the dynamic load factor.

Talking about dynamic buckling compressed structures, the pulse duration should be set close to the period of natural vibrations. Let us check an influence of the pulse duration on deflection—*DLF* curves in the case of the dynamic torque pulse. The results of calculations for the pulse duration changing from $0.5T$ to $2T$ (where T is the period of natural vibrations) are presented in Fig. 9.14. According to the Volmir criterion, the relation between the pulse duration and the critical value of the dynamic load factor is as follows: the longer pulse duration, the lower DLF_{cr}. Taking into account the Budiansky-Hutchinson criterion, the lowest range of DLF_{cr} is obtained for the pulse duration equal to the period of natural vibrations ($T_{\mathrm{p}} = T$). The Ari-Gur and Simonetta criterion cannot be used for all cases presented in Fig. 9.14—for the curve representing the pulse duration equal to half a period of natural vibrations—$T_p = 0.5T$, this criterion cannot be used. Taking into account other cases, the lowest DLF_{cr} according to the Ari-Gur and Simonetta criterion was obtained for $T_p = T$ and it is equal to $DLF_{cr} = 1.6$. Summing up the above, it can be said that the most dangerous period of the pulse duration is the one close to the period of natural vibrations ($T_p = T$) not only for compression but also for the torsional pulse load.

Figure 9.15 presents a relation between the dimensionless deflection and the dynamic load factor for perfect (without any imperfections) girders with different lengths ($l = 100$ mm, 200 mm and 400 mm), subjected to the torsional pulse loading with the pulse duration equal to the period of natural vibrations ($T_p = T$). The critical value of the dynamic load factor estimated on the basis of the above-mentioned criteria is presented in Table 9.4.

Table 9.4 Critical value of *DLF* and the pulse amplitude $(M_t)_{cr}$ for different girder lengths

	Volmir	Budiansky-Hutchinson	Ari-Gur and Simonetta
$l = 100$ mm	$DLF_{cr} = 1$ $(M_a)_{cr} = 4{,}313$ Nm	$DLF_{cr} = 1.4 \div 1.5$ $(M_a)_{cr} = 6{,}038 \div 6{,}470$ Nm	–
$l = 200$ mm	$DLF_{cr} = 1.2$ $(M_a)_{cr} = 3{,}091$ Nm	$DLF_{cr} = 1.4 \div 1.5$ $(M_a)_{cr} = 3{,}606 \div 3{,}864$ Nm	$DLF_{cr} = 1.6$ $(M_a)_{cr} = 4{,}121$ Nm
$l = 400$ mm	$DLF_{cr} = 1.15$ $(M_a)_{cr} = 2{,}420$ Nm	$DLF_{cr} = 1.2 \div 1.3$ $(M_a)_{cr} = 2{,}526 \div 2{,}736$ Nm	$DLF_{cr} = 1.5$ $(M_a)_{cr} = 3{,}158$ Nm

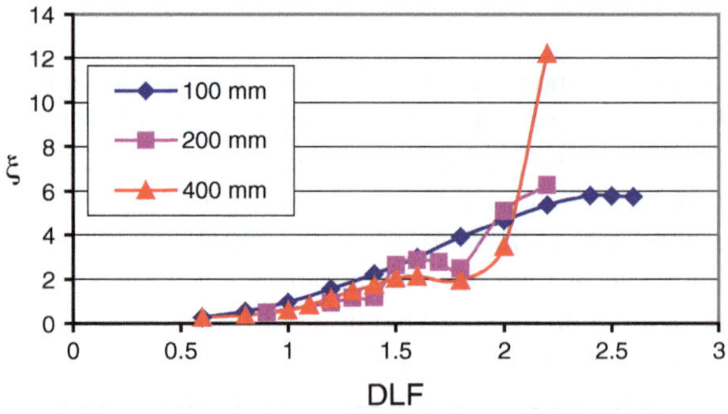

Fig. 9.15 Influence of the length of the girder on the dimensionless displacement versus the dynamic load factor

Comparing the results presented in Fig. 9.15 and in Table 9.4, it can be said that the critical values of *DLF* for different lengths (in the calculated range) are similar. Here, it should be mentioned that *DLF* is a dimensionless value, i.e., the absolute critical amplitude for pulse load is different. Taking it into account, it could be stated that the longer girder, the lower critical amplitude of pulse loading.

References

1. Ari-Gur J, Simonetta SR (1997) Dynamic pulse buckling of rectangular composite plates. Compos B 28B:301–308
2. Czechowski L, Kubiak T (2012) Analysis of thin-walled girders subjected to a pulse torsional torque. J KONES Powertrain Transp 19(1):51–62
3. Hutchinson JW, Budiansky B (1966) Dynamic buckling estimates. AIAA J 4(3):525–530
4. Krolak M, Mania RJ (eds) (2011) Static, dynamic and stability of structures. Vol. 1: Stability of thin-walled plate structures. A series of monographs. Technical University of Lodz Press, Lodz
5. Simitses GJ (1990) Dynamic stability of suddenly loaded structures. Springer Verlag, New York
6. Volmir SA (1972) Nieliniejnaja dinamika plastinok i oboloczek. Science, Moscow
7. SAS IP, Inc (2007) ANSYS 11.1 html online documentation

Chapter 10
Thin-Walled Girders Subjected to Combined Load

This chapter deals with dynamic response of thin-walled girders subjected to combined load (bending and/or twisting). The load was assumed as stepped dynamic pulse of finite duration. Numerical analysis was conducted with finite element method (FEM). Analysis has concentrated on thin-walled girders of length 200 mm and squared cross-section (100 × 100 mm) with 1 mm wall thickness. Different numerical models were compared and investigation of dynamic response of the structure under combined load was performed for different boundary conditions. The critical values of load were determined according to well-known criteria (Volmir, Budiansky-Hutchinson and Ari-Gur-Simonetta criterion).

The dynamic buckling analysis of thin-walled structures subjected to combined load was conducted with the finite element method [1] only. Till now, the analytical-numerical method presented in Chap. 3 is not prepared for solving girders subjected to simultaneous bending moment and torque. In order to obtain results comparable to the real structure, two numerical models were checked. Preparing a discrete model of the analysed girders, special care was taken not to generate stress concentration regions. Different models of load and boundary conditions were considered.

The dynamic buckling of girders subjected to combined pulse loading is analysed [2]. A thin-walled girder of the length $l = 200$ mm and a square cross-section ($a = 100$, $b = 100$ mm) with 1 mm wall thickness was taken into account. The considered structure was subjected to the pulse of bending moment and torsion—the load scheme is shown in Fig. 10.1. It was assumed that the analysed girders were made of an isotropic material (steel) with the following material properties: $E = 2{\cdot}10^5$ MPa and $\nu = 0.3$.

Two taken into consideration FEM models are presented in Fig. 10.2. In order to assure the linearity of loaded edges, two plates of relatively high stiffness were added to the ends of the girder in the first example (Model 1—Fig. 10.2a). In the second example, the same linearity of the loaded edges was obtained with beam elements (Model 2—Fig. 10.2b). These stiff plates and beams were added to the model for better load modelling.

T. Kubiak, *Static and Dynamic Buckling of Thin-Walled Plate Structures*,
DOI: 10.1007/978-3-319-00654-3_10,

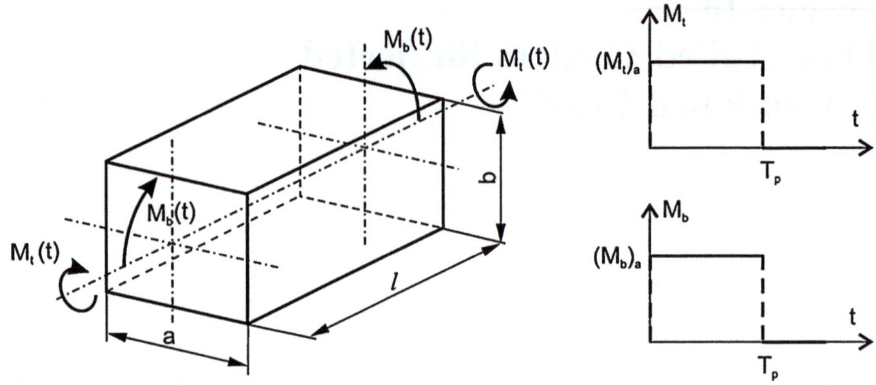

Fig. 10.1 Dimensions and a loading scheme of the girder under analysis

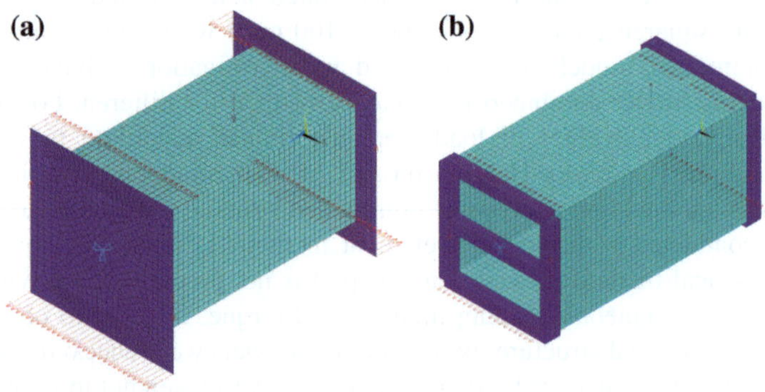

Fig. 10.2 FEM models with a stiff plate (**a**) and with beams (**b**)

The main difficulty is to assume a relation between the bending moment and torque which should be taken in further calculations, i.e., in the dynamic buckling analysis. First, the eigenbuckling problem for different bending to torsional moment ratios was solved. The obtained buckling mode for two models under consideration is presented in Fig. 10.3. The figure shows that both the models yield similar results—the buckling modes were the same for both the models. Critical values of bending and torsional moments are listed in Table 10.1.

Depending on the bending moment and torque relation, combined buckling loads differ for both the models under consideration. The relations of the bending moment versus torque for the combined buckling load are presented in Fig. 10.4.

After the preliminary tests, the beam model was dismissed because of the difficulties with convergence of the dynamic solution. It was caused by a formation of the stress concentration at the corners, which was mainly due to the modelling method of boundary conditions. The pulse duration T_p was obtained on the basis of the modal analysis as equal to the period of natural vibrations and set to

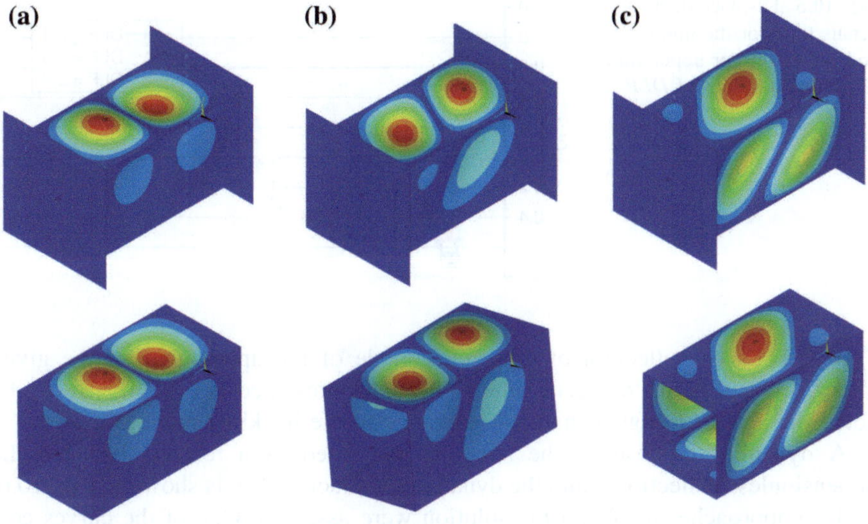

Fig. 10.3 Shape of the local buckling modes: **a** $M_t/M_b = 0$, **b** $M_b/M_t = 1$, **c** $M_b/M_t = 0$

Table 10.1 Critical values of the bending moment and torsion for both the models under consideration

M_b/M_t	Model 1		Model 2	
	M_{bcr} (Nm)	M_{tcr} (Nm)	M_{bcr} (Nm)	M_{tcr} (Nm)
∞	0.725	0	0.732	0
3	0.710	0.237	0.711	0.237
1	0.621	0.621	0.604	0.604
0.333	0.377	1.132	0.348	1.045
0	0	1.512	0	1.335

Fig. 10.4 Comparison of the results obtained for different models

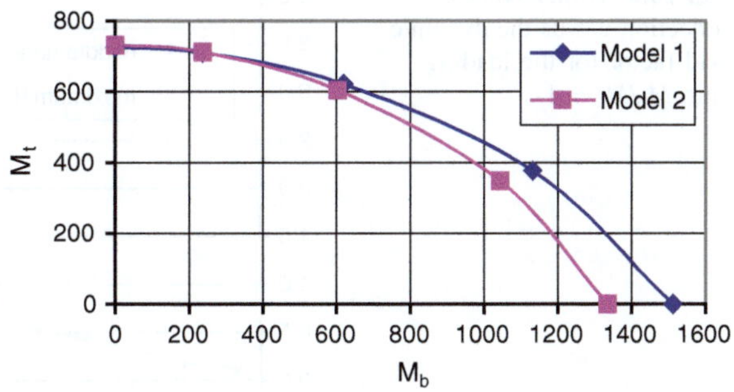

Fig. 10.5 Displacement versus time for the middle node of the girder upper wall for different values of *DLF*

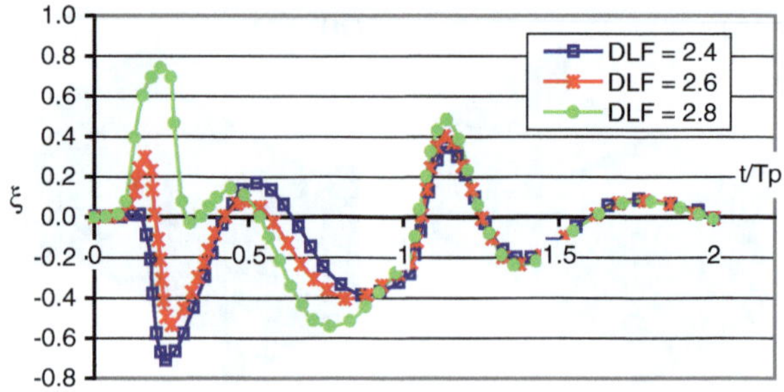

$T_p = 0.003$ sec. Deflection of the middle node of the upper plate of the given girder versus time for different values of *DLF* is presented in Fig. 10.5. It can be noted that for *DLF* higher than 2.4, a change of the buckling mode occurred.

A dynamic response of the structure presented as a relation between the dimensionless deflection ξ and the dynamic load factor *DLF* is shown in Fig. 10.6.

Two approaches to obtain the solution were assumed. One of the curves corresponds to the values of the dimensionless deflection measured in the middle of the upper plate of the girder *(middle node)* and the second one is related to the maximum values of the dimensionless deflection in the whole structure *(maximum deflection)*. The courses of the curves are similar, however, the deflections measured in the middle point do not have the maximum value. Additionally, it should be noticed that the critical value of the dynamic load factor received according to the Volmir criterion [3] differs due to the deflection value obtained from both the curves. The critical values of the dynamic load factor obtained on the basis of different criteria [3–5] are compared in Table 10.2.

Comparing the curves in Fig. 10.6 and the critical values in Table 10.2, one can conclude that deflection depends strictly on the chosen point and it is very difficult to choose the proper criterion to determine DLF_{cr} for combined loads. It is suggested that for combined loads, the dynamic response should be analysed. Because the obtained results are different for each chosen point in the upper wall of the

Fig. 10.6 Dimensionless deflection versus the dynamic load factor for the loading ratio $M_b/M_t = 1$

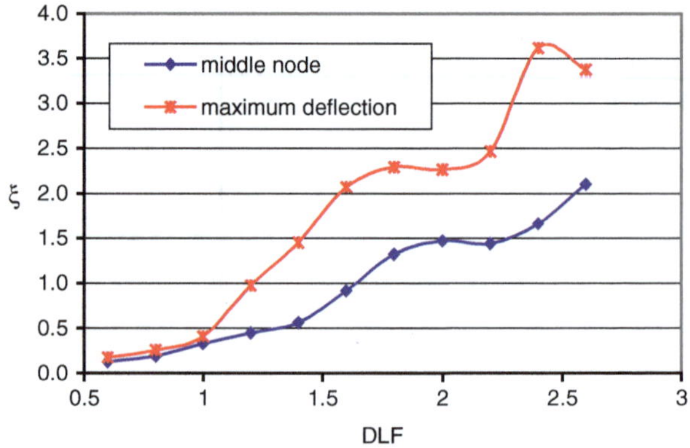

Table 10.2 Critical values of dynamic load factors

	Middle node	Maximum deflection
Volmir criterion	1.7	1.3
Budiansky-Hutchinson criterion	1.5–1.7	1.5–1.7
Ari-Gur-Simonetta criterion	1.8	2

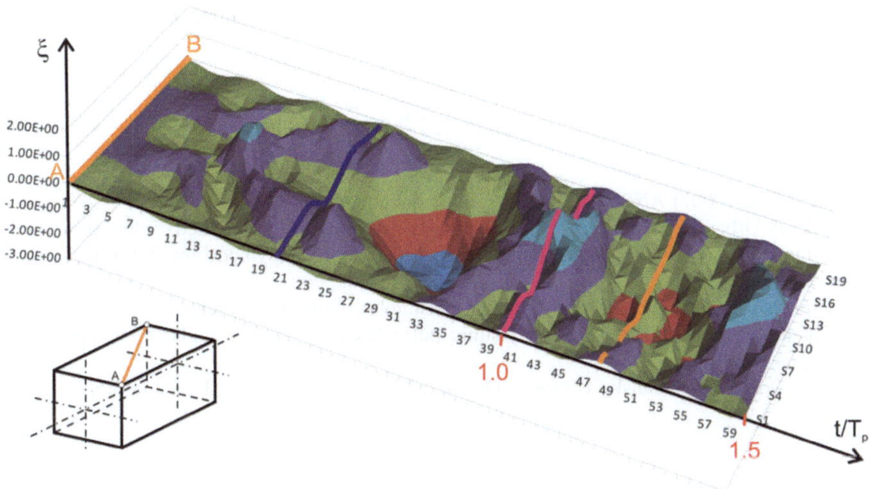

Fig. 10.7 Dimensionless deflection versus time along the wall diagonal

girder, it is proposed to analyse deflection versus time along the diagonal of the girder wall. The proposed way of analysis is presented in Fig. 10.7. Such a representation of deflection versus time allows one to find the maximal deflection as a function of time on the wall under consideration.

Analysing the results presented in Fig. 10.7, a change in the mode shape with an increase in the dynamic load factor can be found. This phenomenon is shown in Fig. 10.8, where deflection with respect to the position along the diagonal of the

Fig. 10.8 Dimensionless deflection along the wall diagonal

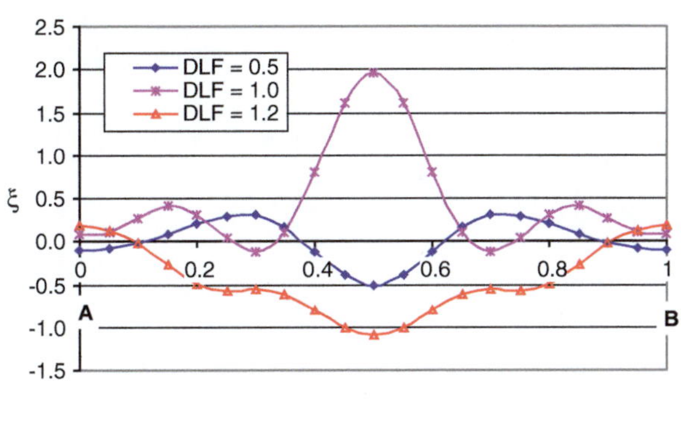

upper wall for different *DLF* values is presented. A change in the mode shape appears between $DLF = 1$ and $DLF = 1.2$.

To conclude, one should emphasize the fact that the behaviour of structures (in the presented case—the girder) subjected to the combined pulse load is different than for the case when structures are subjected to a single type of load. Analysing the results of calculations presented above, it can be said that a criterion checking a change in the displacement mode, i.e., the Ari-Gur and Simonetta criterion [4], is the most suitable criterion for structures under combined pulse loading.

References

1. SAS IP, Inc (2007) ANSYS 11.1 html online documentation
2. Wluka P, Urbaniak M, Kubiak T (2012) Dynamic response of thin-walled girders subjected to combined load. In: Proceedings of stability of structures 13-th symposium, Zakopane, Poland, pp 677–686
3. Volmir SA (1972) Nieliniejnaja dinamika plastinok i oboloczek. Science, Moscow
4. Ari-Gur J, Simonetta SR (1997) Dynamic pulse buckling of rectangular composite plates. Compos B 28B:301–308
5. Hutchinson JW, Budiansky B (1966) Dynamic buckling estimates. AIAA J 4–3:525–530

MIX
Papier aus verantwortungsvollen Quellen
Paper from responsible sources
FSC® C105338

If you have any concerns about our products,
you can contact us on
ProductSafety@springernature.com

In case Publisher is established outside the EU,
the EU authorized representative is:
Springer Nature Customer Service Center GmbH
Europaplatz 3, 69115 Heidelberg, Germany

Printed by Libri Plureos GmbH
in Hamburg, Germany